An Introduction to Tourism and Anthropology

D0224375

Tourism is more than a collection of business transactions, a process, or set of impacts. It is a complex assortment of systems that includes economic, built and natural environments, ownership patterns, relationships between generating and receiving countries, and the relationship between the locale in which tourism takes place and the wider society.

An Introduction to Tourism and Anthropology explains how anthropology is the window through which tourism dynamics may be properly analysed and evaluated. Starting with an overview of the development of anthropology as a social science, the author moves on to examine:

- the definition and characteristics of tourism
- definitions and typologies of tourists
- tourism and culture
- key themes and writers in the anthropology of tourism
- issues in the anthropology of tourism
- globalisation, tourism and anthropology
- the relationship between development and underdevelopment theories and the anthropology of tourism

Featuring a wealth of international examples, figures, key ideas, guides for further reading and end of chapter questions, this comprehensive book examines the important issues surrounding tourism's human impact and reveals that the dynamic relationship between tourism and culture is not necessarily a 'bad thing'.

Peter M. Burns is Head of Tourism and Leisure at the University of Luton.

Front cover illustration

Land diving ceremony on the island of Pentecost (Vanuatu).

This complex ceremony has been performed by men and boys of Pentecost island for hundreds of years. Its purpose has been twofold. First, to prove manhood and to provide an initiation focus for boys, and secondly, through the diver's head brushing the soil at the end of the jump, to symbolically fertilise the earth for the next season's yam crop. Tourism, and the desire to generate cash income for village projects, has given a third reason. Perhaps the dive now symbolically re-fertilises next year's crop of tourists.

Each diver selects their own vines, cuts them and binds the end with damp leaves so that they will not dry out. The length is decided according to where the person will dive from and their weight. This has to be done very carefully as the desired effect is to dive and just touch the earth with the head. The tower, constructed out of branches and saplings, can be as tall as twenty five metres. It symbolises the shape of a man, and the diving platforms are placed strategically to coincide with 'knees', 'waist', 'shoulders' etc. The person who organises the practical elements of the dive (the ceremonial aspects remain in the hands of the chiefs) dives from the 'head' and the dive is named after him. The dance is accompanied by chanting and singing from the village women. This photograph was taken by the author in May 1988, and was the first dive to be organised jointly by the National Tourist Office of Vanuatu, Tour Vanuatu (a tour operator owned by the government) and the chiefs of southern Pentecost. The ceremony was conducted entirely in accordance with tradition, although we found out later that some of the money was allegedly diverted to a local politician and not the promised village project. The paradox for any aspiring anthropologist would be trying to resolve the extent to which this event was 'authentic'. The tourists I talked to on the trip back all thought the whole event was better than their wildest expectations: for them (and me) there was no doubt about the value of this amazing cultural experience.

An Introduction to Tourism and Anthropology

Peter M. Burns

First published 1999 by Routledge
11 New Fetter Lane, London EC4P 4EE

Simultaneously published in the USA and Canada
by Routledge
29 West 35th Street, New York, NY 10001

Typeset in Sabon by Keystroke, Jacaranda Lodge, Wolverhampton
Printed and bound in Great Britain by Biddles Ltd, Guildford and King's Lynn

British Library Cataloguing in Publication Data
A catalogue record for this book is available from the British Library

Library of Congress Cataloging in Publication Data
Burns, Peter (Peter M.)
    An introduction to tourism and anthropology / Peter M. Burns,
      p. cm.
    Includes bibliographical references and index.
    1. Anthropology. 2. Tourist trade. I. Title.
    GN27.B95 1999
    301—dc21   98–37417

ISBN 0–415–18626–9 (hbk)
ISBN 0–415–18627–7 (pbk)

Contents

Figures

Tables

Preface

An increasing number of students are writing essays, dissertations and theses based on the social and cultural impacts of tourism. Two dangers arise. First, from the uninformed lecturer who may have a background rooted in quantitative research and/or business who may level the criticism that these types of essay are 'unscientific' and emotional rather than rational and empirical. The second danger is that they might be right! Students may in fact take a very lightweight, 'surface' look at a particular aspect of tourism and base their work on socially constructed interpretation rather than objective analysis.

This book is intended to illustrate how qualitative research, especially anthropology, has been brought to bear on the study of tourists and tourism. Anthropology is about fieldwork, recording of information, accounts of human culture and social organisation. Social anthropology in particular puts its interest in the close and detailed study of particular communities or sub-cultures. In this sense it differs from sociology which is usually more concerned with analysing social trends in society, often using statistical methods. We can see then that anthropology is a science *par excellence* to use for the qualitative study of tourists and tourism.

In writing this book, my hope is to introduce students to a number of key ideas (linked to their authors), explain what they are about, and direct them to the original sources. Thus if students find the work of Graham Dann or Valene Smith interesting, then they should go to the original sources! There is really nothing to replace reading an author's own words, even if there are texts, such as the present one, which offer an explanatory framework. In this sense, my book should be seen as a beginning rather than an end.

Peter Burns, University of Luton
July, 1998

Acknowledgements

A number of people read early drafts of this book and were kind enough to comment on the content, style and approach. They also encouraged me in my work. In particular I am grateful to: Emma Heard (MAITP student, class of '97); Professor Mike Hitchcock; Dr Julie Scott of the Eastern Mediterranean University; students in the class of 1998 'Tourism Systems' module at University of North London; Sarah Macmillan, final year student in the class of 1998 BA (Hons) Tourism; Raoul Bianchi of Derby University; Kiran Kalsi, University of North London; John Coventon (to whom I still owe a bottle of wine); and Oliver Bennett of Touche Deloitte with whom I've enjoyed an ongoing debate about the 'use' of academics in tourism consultancy! I am also grateful to Lierni Lizaso Urrutia (BA (Hons) Tourism class of 1993) who did much of the original literature research. While Professor Chris Cooper has not been directly involved in this particular project, his general faith in my ideas has proved invaluable. Casey Mein, assistant editor at Routledge, also deserves a vote of thanks for maintaining her enthusiasm for the project even though it dragged on a little longer than it should have done! Finally, thanks to Andrew Holden for being a good and consistent friend, and of course Professor Tom Selwyn for being my mentor.

I don't know how to go about thanking the people I have met in my travels (work and pleasure) who have answered my questions, listened to my comments and enriched my knowledge and understanding, I guess all I can say is that you remain in my thoughts even when my promises of 'going back' are broken.

Part I

Anthropology, tourism and tourists

Part I consists of four chapters. Chapter 1 gives an overall context to the book by briefly examining the development and history of anthropology as a science. The chapter ends by discussing some of the current trends in anthropology. Chapter 2 reviews some of the attempts at theorising tourism and explains how it is useful to think of tourism as a *system* rather than simply as a *process* or set of *phenomena*. Following on from this, Chapter 3 looks at ways in which tourists have been defined and discussed in the academic world, and discusses some of the more complex motivations that cause people to become tourists, while Chapter 4 threads this together by examining the building block of society: culture.

1 Anthropology

Plate 1: This picture follows on from the front cover illustration. It shows a group of Pentecost islander men and boys wearing traditional dress (the penis sheath or *namba*) and preparing themselves for the land dive. A picture of tribal peoples is one readily associated with anthropology, although as the chapter will reveal, anthropology is about much more than far away, remote people. It is now used to help understand sectors of society and institutions in Western culture. The picture could easily have been of new age travellers, elderly people in a nursing home or farmers at a cattle market.

Overview, aims and learning outcomes

Anthropology is the study of humanity. It seeks to understand and explain how human societies work. This chapter traces some of how anthropology has evolved since its beginnings in the mid nineteenth century. The key schools of thought and approaches that arise from conducting anthropological research in a world of escalating complexity are discussed. The specific aims of the chapter are to:

- chart the evolution of social anthropology;

- discuss the main approaches to anthropology; (and)

- identify some of the future directions that anthropology might take.

After reading this chapter you should be able to:

- differentiate between the key approaches to anthropology;

- evaluate the implications arising from these approaches; (and)

- describe some of the ways in which anthropology is shaping up to meet future challenges

Introduction

Living in London, there are certain rules by which I interact with my fellow Londoners. When buying food in a supermarket, I know what most of the foods are and that I have to line up at a queue in order to purchase them; I will probably not engage in conversation with other shoppers whom I do not know. When travelling on the London Underground, I know not to talk to my fellow passengers, nor even, for the most part, to look them in the eye. Thus, when I went to live in Fiji with my family, we were overwhelmed with friendliness as people said good morning to us in shops and on the streets. Travelling on the buses with our young children, we were amazed when our babe-in-arms was taken from us and passed round the bus so that the Fijian women could get a close look and kiss the baby. My family and I had to learn how to say yes to offers of food (so as not to offend), not to admire material things too openly (or we would find the item wrapped up and given to us) and teach our children not to ask for things (as they would inevitably be given them).

The social rules were different, and we had to learn to operate within them. We suffered a mild sort of 'culture shock' albeit a particularly pleasant one. On reflection, it is not difficult to see why, at one level, things are different in London and Fiji. On the one hand, London has a population of 6.8 million, the whole of Fiji has a population of about 730,000. London is a frenetic, post-industrial mega-city while Fiji remains largely rural. In addition to learning 'how to behave' in the streets, on the buses, in the shops, we also had to learn about the complex culture that frames relationships between indigenous Fijians and the Indian population bought to Fiji as part of Britain's colonial policy. That wasn't the end! There was also the question of the 'expat' culture. At the time there were about two hundred expatriate families from Australia, New Zealand, Britain etc. This community had its own culture with both formal and informal rules governing behaviour codes and leisure patterns.

It is these complexities, social interactions, rules, conflict resolution, food habits, attitudes towards strangers, belief systems and a host of other elements that make up culture that holds the attention of anthropologists.

The beginnings of anthropology

Anthropology as we think of it today emerged during the mid nineteenth century: a very interesting and lively period of history.[1] At this time (more-or-less coinciding with the Victorian era) at least three major themes were present in the thinking of educated Western society. These were colonialism, missionary societies, and Darwinism. Their proximities to each other are shown in Figure 1.1.

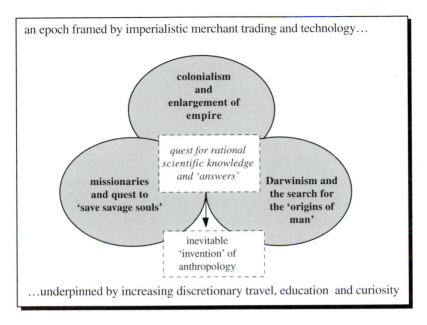

an epoch framed by imperialistic merchant trading and technology...

colonialism
and
enlargement of
empire

*quest for rational
scientific knowledge
and 'answers'*

missionaries
and quest to
'save savage souls'

Darwinism and
the search for
the 'origins of
man'

inevitable
'invention' of
anthropology

...underpinned by increasing discretionary travel, education and curiosity

Figure 1.1: Issues in mid nineteenth century Western society

The three main elements shown in the model above were all interconnected. The missionaries were bringing God and 'civilisation' to the native populations of the colonial world. In their earliest days (say, the mid to late eighteenth century, roughly during the Age of Enlightenment[2]), these missionaries might have carried the 'romantic' idea that the societies being 'discovered' through imperialism and its political tool, colonialism, were the archetypal Rousseau-esque[3] 'noble savages' unspoilt and living a simple life, just as God intended. In later times, as the nineteenth century arrived, they may well have held the view that such peoples were 'primitives' low down in the hierarchy of society who needed to be civilised. The colonial officials were, very likely, convinced of their racial superiority over their subjects (a reflection of the unreconstructed attitudes at that time) and popular misinterpretations of Darwin's theory strengthened the idea of a unilinear evolution of species where only the fittest survived (the biological evolutionist school of thought). This was not quite what Darwin, who emphasised the randomness of nature, intended. Given this powerful mix of cultural politics, the sense of superiority in Western society brought about by industrialisation (i.e. bringing nature under control) it is hardly surprising that a scientific response to the study of the human species emerged. Figure 1.1 is one interpretation of the state of things at that time, a version of events that reflects current mainstream thinking on the history of anthropology. It should not be taken as the only way of looking at the socio-political structures that existed. Edward Said, a Palestinian-born intellectual and Professor of English and Comparative Literature at Columbia University, describes this era as having four elements:

1   an expansionist one, packed with imaginary utopias, and an increasing sense of European cultural strength;

2   'a more knowledgeable attitude towards the alien and exotic was abetted not only by travellers and explorers but also by . . . confront[ing] the Orient's peculiarities with some detachment and with *some attempt at dealing directly with Oriental source material . . . [letting] Muslim commentators on the sacred text [the Koran] speak for themselves* (1978:117, italics added)';

3   'sympathy' or 'Popular Orientalism' where 'true' knowledge about Other could only be gained by suspending all prejudice whereupon one might 'see hidden elements of kinship between [oneself and the Orient' (1978:118);

4   finally, 'the impulse to classify nature and man into types . . . there [was] everywhere a similar penchant for dramatizing general features, for reducing vast numbers of objects to a smaller number of orderable and describable *types*' (1978:119, italics in original).

These four points are important because it is an early introduction to some of the problem areas we shall be examining later in this book.

Table 1.1: Some important dates for early anthropology

1839	Société Ethnologique de Paris founded in France
1843	Ethnological Society of London formed
1859	Charles Darwin publishes *On the Origin of Species*
1871	Royal Anthropological Society of Great Britain and Ireland founded
1871	E. B. Tylor publishes *Primitive Culture*
1871	Sir John Lubbock publishes *Origin of Civilization*
1871–82	Russian naturalist Nikolai Milouho-Maclay conducts first recognisable *extended* fieldwork studying the people of Madang in New Guinea
1876–96	Herbert Spencer (1820–1903) publishes *The Principles of Sociology* (three volumes) which suggest that human society can be compared to a biological organism (i.e. in both cases all the parts have to be studied before the 'whole' can be understood)
1877	Lewis Henry Morgan (1818–81, United States) publishes *Ancient Society* about kinship, family and property among Iroquois Indians
1879	University of Rochester (USA) commences anthropology courses
1879	Bureau of American Ethnography established at the Smithsonian Institution
1884	Sir Edward Tylor appointed Reader in anthropology at Oxford University
1888	Harvard and Clark Universities (USA) start departments of anthropology
1896	Franz Boas (1858–1942) (German, based at Columbia University in the United States) published *The Limitations of the Comparative Method of Anthropology* criticising the evolutionists whom he saw as racist
1898–9	Cambridge University expedition to Torres Straits (between Australia and New Guinea) led by Alfred Haddon (a zoologist) and included W.H.R. Rivers (a psychologist)
1908	First British chair (professorship) of anthropology for Sir James Frazer (1854–1941) at Liverpool University where he wrote *The Golden Bough*
1915–18	Bronislaw Malinowski (1884–1942) extended fieldwork in the Trobriand Islands; participant observation with focus on culture
	Radcliffe-Brown (1881–1955) focus on data gathering through fieldwork especially concerning social relations/society/ maintenance of social structure

Table 1.1 is a chronology which traces some of the key dates important for early anthropology. The purpose of compiling this list is simply to put into chronological order some of the names that are frequently referred to so that you can see what events followed what. In their own way, each of the names and events listed made important contributions to the development of anthropology. However, if we had to choose those with a particular influence on the shape of anthropology, Boas, Haddon, Malinowski and Radcliffe-Brown are of particular interest because of their attempts to develop systematic ways of collecting data through personal observation, and their emphasis on explaining customs and beliefs within the context of their social and cultural settings. They set the idea of *fieldwork*[4] as being of indispensable for anthropology.

In this way, with their emphasis on participant observation[5] and extended periods in the field, they differentiated anthropology from its early 'armchair' roots. For example, Frazer's seminal work on the evolutionary nature of human life and the origins and characteristics of religion, *The Golden Bough* (published in thirteen volumes from 1890 on), was not based on empirical research or field observations, but rather on academic reflection, sometimes rather simplistically referred to as 'armchair anthropology'.

To place the dates listed in Table 1.1 in a slightly more reflective context, Figure 1.2 shows the main themes of early anthropology and some of the philosophies behind them. The words 'deterministic' and 'hierarchical' in this context mean a belief in a natural, unchangeable, or even divine order of things, for example the role of women in society being somehow 'pre-ordained' or that some cultures are 'naturally' better (more important) than others. In general, the view of the world expressed in Figure 1.3 could be described as both deterministic and hierarchical.

During the mid to latter part of the twentieth century the key theoretical framework has been *structuralism*[6] most closely associated with the Belgium anthropologist Claude Lévi-Strauss. Central ideas that helped him develop his theory of structuralism were that:

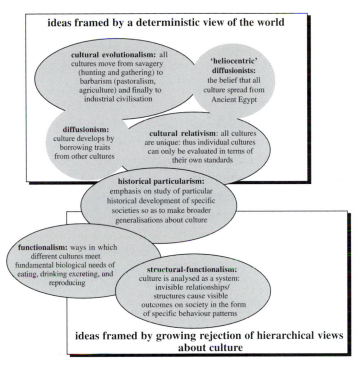

Figure 1.2: Themes in early anthropology

- humans endlessly rearrange and categorise and classify the world about them;
- all human minds, and therefore all thought processes, are basically alike throughout the world; (and)
- thought processes are structured into contrasting pairs of symbolic dualisms (opposites, bi-polarities, or symbolic oppositions) starting with 'dark : light' through to 'male : female'; and 'self : other' ending, ultimately, with 'birth : death.'[7]

Lévi-Strauss' work is about the design of cultures. So, as well as looking at social structure to account for the way in which people in a society behave, he looked for *deeper meaning*. He did this not so much by looking at immediately observable behaviours or material culture, but by searching for the relationship between these social structures and the deeper mental structures (such as language, kinship, and myths) that underpin society.

Lévi-Strauss believed that there is no difference in the application of thought processes within a society governed by 'primitive' mythology and one governed by a 'scientific' framework: 'the kind of logic in mythical thought is as rigorous as that of modern science, and . . . the difference lies, not in the quality of the intellectual process, but in the nature of things to which it is applied' (Lévi-Strauss, 1963:230).

Thus, in some societies the intellectual world has expanded enormously (mainly through technology) and so the products of that mind have been transformed, as Keesing and Keesing describe it: 'Men[8] everywhere are plagued intellectually by the contradictions of existence – by death; by man's dual character, as part of nature yet transformed by culture . . . the realm of myth is used above all to tinker endlessly with these contradictions' (1971:311).

Given that being on holiday is the opposite of being at work, it can be seen that much of Lévi-Strauss' work is relevant to the deep analysis of tourism. MacCannell suggests that, 'In traditional society, man could not survive unless he oriented his behavior in a "we are good – they are bad" framework.' (1976:40). The idea of *opposite states of being* is clearly important to understanding tourism, as are his ideas about the role of myths. If we generally agree with the proposition that the term 'myth' refers to something that is not, in a literal sense, true[9] then we probably agree that at a most basic and obvious level, *myths function as cultural history*. They legitimise religious belief and help with the continuity of the existing social order.[10] However, myths are more than mere projections or reflections of an existing social order. Keesing and Keesing argue that:

> Lévi-Strauss is seeking to explicate the universal [i.e. not specific to one culture] workings of the human mind by looking at varied cultural forms as its artefacts. The realm of myth is crucial in this enterprise because here human thought

has its widest freedom. Not every imaginable form of marriage, house style, or residence pattern is actually found – there are too many constraints, too many possibilities that are unworkable for ecological, technical, or purely physical reasons. But man can *think* all these possibilities, and in myth his thoughts have freest reign.

(1971:311)

In the urban, post-industrial environment that generates most of the world's tourists, part of that world is a disconnection with nature and spirituality, the abiding myth of such an existence is that of *freedom*. It is travel and tourism that keeps this particular myth alive and continues with the idea that travel is, as the US Travel Administration claims, 'The Perfect Freedom'.

Developments in anthropology

The themes that unify anthropology (Haviland, 1990) are shown in Figure 1.3; they make for a distinct way of studying human life.

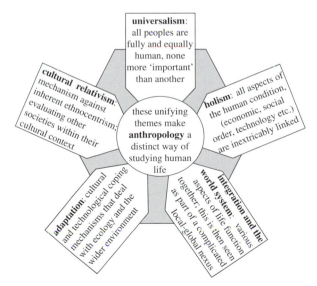

Figure 1.3: Themes that unify cultural anthropology
Source: after Howard 1996: 3–8

These themes were not always given equal weighting. Keesing and Keesing reflect on the rise of colonialism and the accompanying 'cultural aggression' in the form of missionaries, suppression of culture and imposed social change. They continue:

> Colonial administrations varied in their treatment of tribal peoples, from ruthless suppression to paternalistic protectionism and sympathy to local custom . . . Yet the cynic can find much evidence that even now [in more enlightened times] when the stakes are sufficiently high, native peoples who stand in the way of dam, uranium, copper, or airfield will be brushed aside as efficiently as ever.
>
> (1971:359)

While it must be clearly acknowledged that efficiently brushing people aside for development is not the sole prerogative of colonial administrations, it was they who put a sophisticated administrative gloss on the matter. Here are the same authors again:

> However oppressive or well-meaning colonial administrations, missionaries, and others have been, they have made one thing uniformly clear to the colonized peoples: European superiority. 'The native' was a creature apart, removed by his skin color, his cultural backwardness and ignorance, and the other unfortunate circumstance of his birth, from the grace open to Europeans. The symbolic relegation of 'the native' to a position midway between animal and Western man was made very clear on every side . . . the stigma of their birth symbolically remained even [if] they were trained to rise above and reject their traditional ways of life.
>
> (1971:360)

The immediate post-colonial period (roughly the decade of the late-1940s to the middle of the 1960s) were paradoxical times for anthropology. On the one hand, excellent ethnographic fieldwork and theoretical developments continued (for example Edmund Leach's work in Burma, Lévi-Strauss' work on linguistics, myth, social organisation etc.) while on the other hand, some assert that the role of *applied anthropology*[11] languished. For example, Howard observed that many of its formative theories were found to be wanting, and that:

> The destruction of the colonial order following World War II left anthropology, particularly applied anthropologists, open to attack. In many parts of the world, nationalist leaders who had replaced colonial administrators identified anthropologists with the former colonial regimes, viewing anthropologists as agents of colonial repression.
>
> (1996:400)

It would not be quite right to make too much of the links between anthropology and colonialism, but they do exist, even if only in a small way. However, in the face of certain paradoxes, the subject has found new directions. For the purpose of the present volume, prime among these has been anthropological involvement with Development Studies (Third World[12] issues) where anthropologists have brought new insights to the impact of development projects. Anthropologists have been acting as go-between in discussions between local people and central (or other) planners or indeed foreign developers. Here is Howard again:

> Typically, the local population distrusts the planners. Such distrust may be warranted because planners often lack knowledge about the local culture and the specific needs and goals of the local population. To bridge the gap between the people and the planners, the anthropologist must (1) identify and explain the local decision-making processes and (2) help both residents and planners adapt local structures to better meet development needs as expressed by the residents.
>
> (1996:401)

One possible interpretation of this is that anthropology remains what it started out as: whites studying 'Other'. This is not Howard's intention, but time and time again one gets the sense of anthropologists as plantocrats (a plantation-owning ruling élite), guarding their (intellectual) property with an eagle eye.

In the case of Britain, anthropologists have long had official contacts with the former Overseas Development Administration (ODA) and this continues with the newly formed Department for International Development (DfID).

The maturing of applied anthropology through its role in sustainable development, women in development and advocacy[13] has probably provided the biggest impetus for the profession. Anthropologists are now working in corporate businesses helping the understanding of the internal organisational culture (Hofstede, 1991). The overall aim is to improve the effectiveness of planning and decision-making just in the way Howard discussed in the above quote. The medical profession has taken on anthropologists to create a deeper understanding of the cultural implications of ageing, drug addiction and the development of risk-reducing strategies for AIDS sufferers. In each of these cases, anthro-pologists act as culture-brokers, clarifying issues and positions between officialdom and the wider population. The notion of 'anthropology-at-home' is of increasing importance with American anthropologists trying to help growing despair, disaffection and crime in inner-cities. In the case of tourism, anthropologists have shifted ground somewhat from reporting problems through ethnographic field work to becoming involved in the process of conflict resolution.

One approach to anthropology is described by Keesing and Keesing in an American college text where, in describing the need for a theoretical framework to shape the discussion on changes to social structures resulting from ecological adaptation and in response to humankind 'as creator and manipulator of rules' (1971:232), they suggest, perhaps based on Gluckman's (1955) *neofunctionalist*[14] ideas on the role of conflict in culture) that society could be viewed as containing a series of battle-lines 'that are always being fought or negotiated'. The metaphor of war is interesting given that Keesing and Keesing's original work was done at the height of the Cold War. This remains an excellent summary of what much of any society is about and is shown in Table 1.2.

Table 1.2: Tensions that shape society

the status of men vs. women	as represented in the division of labour, sleeping arrangements, property rights, lines of descent, ritual participation, rules of sexual conduct etc.
ties between husband and wife	relative importance of husband and wife vs. the tie between brother and sister (the competing rights over a woman of her husband and brother)
collective rights vs. individual rights	unity of a clan in corporate action as opposed to family rivalry and separation; the strength of blood ties from a group of people who are descended from a common ancestor (lineage) as against kinship (networks of relationships including marriage)
conflict over resources vs. common goals of 'peace'	conflict over access to resources such as land, women, property, prestige vs. unity in the pursuit of common goals such as peace, collective security, joint enterprise
unity of peers ('age-mates')	unity of age-mates in different kin-groups vs. the ties of kin group membership that separate them
generation conflict	where a younger generation may 'fight' an older generation for power, authority or resources

Source: after Keesing and Keesing, 1971

Keesing and Keesing finish their description of this approach to the study of society by stating that social structure and the broader environment has a relationship which is indirect, constructed by a variety of ideologies and negotiation, the realisation of which, as they say:

enables us to break away from 'equilibrium models' and views of social structures as 'frozen' that hamper our understanding of change, and it connects the abstract formal structure of 'the system' with the dynamics of social process and individual psychology.

(1971:233)

The reference to 'equilibrium' models of society is an allusion to the somewhat outmoded school of thought that there is a steady and 'normal' state for society to return to if it is somehow disturbed.

In general, the 'study of society' tends to be called *social anthropology* by the British anthropologists and *cultural anthropology* by American academics; there is no significant difference between the two, but approaches will differ on each side of the Atlantic. Haviland, an American professor of anthropology describes the themes in cultural anthropology in a way shown in Figure 1.4:

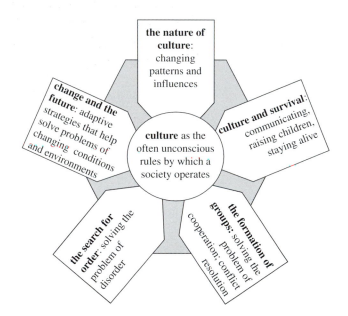

Figure 1.4: Key themes in cultural anthropology
Source: after Haviland, 1990

In addition to the ideas of cultural or social anthropology being applied to different societies, they can also be applied to businesses (cf. Hofstede 1991, above). All organisations have culture. Think of belonging to a group of friends that 'just hang out together' or the scouts/guides associations/gardening or swimming clubs etc. Each will have its set of rules that enable behaviours, actions and attitudes to be 'judged' as appropriate or inappropriate by other members of that group. The same applies to a business organisation which is sometimes described as having a corporate culture. Figure 1.5 shows the elements of such a culture:

Figure 1.5: Elements of corporate culture

These elements of corporate culture are important when analysing the formal and informal power relations within a company, internal and external communications and ways in which changing internal and external environments affect the various parts of it. Understanding how a company works at a deep level is essential in implementing change most effectively.

All aspects of our lives are affected by a series of ever changing cultures that we encounter from time to time. These cultures are influences upon us and we in turn influence them. Cultures are thus influenced by both internal and external factors, willingly and unwillingly. Chapter 4 explores the theme of culture in more depth.

Key ideas

For this chapter, these are best explained by reiterating the key areas of interest that have historically combined together to 'make' anthropology. These can be summarised as:

- a growing sense that there was something of great importance to be learned from making comparative studies between different societies;

- myths are not only to be found in pre-industrial societies, they are also present in industrial and post-industrial society and are just as important in acting as a component of social cohesion; (and)

- an awareness that social problems in all societies (including those located in industrial economies) could be better addressed through an understanding of the deeper structures that underpin such societies.

Questions

1 Does *society* evolve in the same way that Darwin meant for non-human biological beings?

2 What are the main arguments for and against the idea that anthropologists were a tool for colonialism.

3 Discuss the different ways in which the concept of 'culture' may be defined.

4 Describe the culture of an organisation with which you are familiar.

5   Write an anthropological account of your family.

6   What are the ways in which an anthropological study could contribute to an understanding of so-called 'welfare-dependency' in a post-industrial society?

Key readings

To get a real understanding of anthropology rather than the thumbnail sketch presented above, it is essential to read one or more of the classic ethnographies. Malinowski's *Argonauts of the Western Pacific* comes to mind, as does Radcliffe-Brown's *The Andaman Islanders*. For a serious introduction to anthropology as a subject, either Angela Cheater's *Social Anthropology: an Alternative Introduction* or Kuper's *Anthropology and Anthropologists: the Modern British School* are both very useful. Finally, for a populist tale of modern field research in Africa, Nigel Barley's *The Innocent Anthropologist* provides a lively read although some anthropologists find it patronising at the expense of tribal peoples (at the same level as Peter Mayle's *A Year in Provence*).

For an idea of how anthropology is helping with development, see Mark Hobart's edited collection *An Anthropological Critique of Development* (especially Professor Douawe van der Ploeg's chapter on 'Potatoes and Knowledge') and Johan Pottier's volume called *Practising Development* particularly Susan Hutson and Mark Liddiard's essay on 'Agencies and Young People: Runaways and Young Homeless in Wales' and Bill Garber and Penny Jenden's chapter on 'Anthropologists or Anthropology? The Band Aid Perspective on Development Projects'.

2   Tourism

Plate 2: Probably one of the world's most obvious tourist icons: the Eiffel Tower. This edifice symbolises both the city of Paris and, in a sense, tourism itself. The tower serves no particular function apart from its touristic one. It is almost inconceivable to visit Paris without visiting this monument. The question an anthropologist might ask is 'What is it about this tower that makes it draw in tourists who seem to hold it such high regard?' Answers might be found through semiotics (the study of signs and symbolic meanings). Dean MacCannell has much to say on this topic.

Overview, aims and learning outcomes

This chapter examines a number of definitions and ideas about tourism that help us to understand the extent of its complexities. The chapter is not intended to be a full-scale exploration, but rather to provide a tourism context for this book. Specifically, this chapter aims to:

- define tourism and discuss it from a systems perspective;

- identify the main approaches to the analysis of tourism; (and)

- provide an insight into some of the major issues of contemporary tourism.

After reading this chapter you should be able to:

- differentiate between tourism as 'business' and tourism as 'phenomenon' approaches;

- evaluate the rationale for describing tourism as a system; (and)

- describe how tourism is a multi-disciplinary subject.

Introduction

Van Harssel sets a useful context for tourism with the following:

> It is difficult, and perhaps misleading, to generalize about tourism and tourists. We lack a commonly accepted definition of tourism partially because of the complexity of tourist activity and partially because different interests are concerned with different aspects of tourist activity.
>
> (1994:3)

Even with the issue of definition about its precise meaning remaining unresolved, it is generally agreed that there are four primary elements of tourism. These are i) travel demand, ii) tourism intermediaries, iii) destination influences, all of which lead

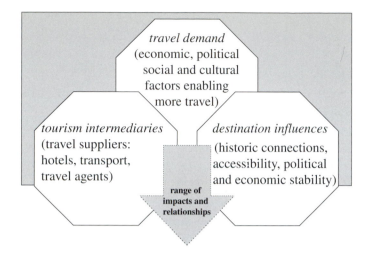

Figure 2.1:   Primary elements of tourism

to iv) a range of impacts. Figure 2.1 shows these elements linked together as a schema.

In addition to the model shown in Figure 2.1, Smith proposed an interesting alternative perspective by describing tourism as a social practice:

> The phenomenon of tourism occurs only when three elements – temporary leisure + disposable income + travel ethic – simultaneously occur. It is the sanctioning of travel within a culture that converts the use of time and resources into spatial or geographical social mobility. If travel is not deemed culturally appropriate, then time and resources may be channelled elsewhere.
>
> (1981:475)

There are many other ways in which tourism can be studied. Put simply, and following on from Buck's early idea (1978), tourism can be seen through opposing schools of thought: tourism as business versus tourism as problem (or set of phenomena). While this is a useful reminder of the tensions that exist between various

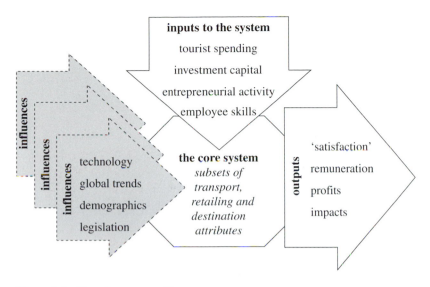

Figure 2.2: Tourism system (i)
Source: after Laws, 1991

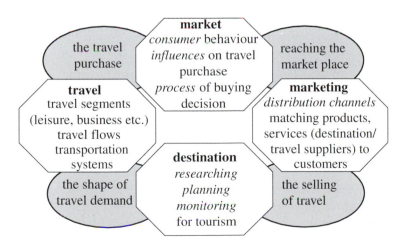

Figure 2.3: Tourism system (ii)
Source: after Mill and Morrison, 1985

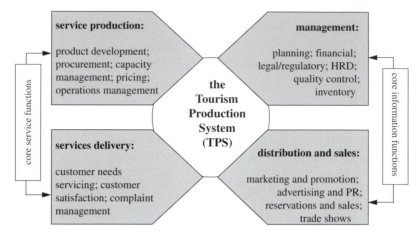

Figure 2.4:   Tourism system (iii)
Source: after Poon, 1993

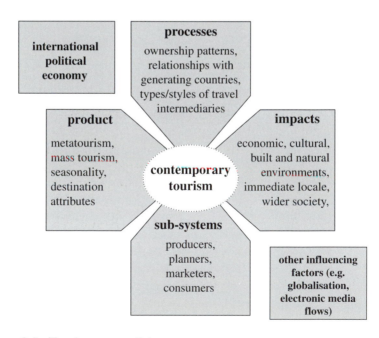

Figure 2.5:   Tourism system (iv)
Source: after Burns and Holden, 1995

groups of people who write and think about tourism, neither method is productive if taken in isolation. I would suggest that one particularly effective approach is through seeing tourism as a system or set of sub-systems. Figures 2.2 to 2.5 show a range of tourism systems, each one having its own perspective but remaining useful in one way or another.

The advantage of a systems approach is that tourism is not automatically seen in isolation from its political, natural, economic or social environments (although this depends on the systems model being used). It emphasises the inter-connectedness between one part of a system and another. This encourages multi-disciplinary thinking which, given tourism's complexities, is essential to deepen our understanding of it. Having an understanding of the tourism system at a particular destination will enable a far better grasp of the *processes* of tourism, thus moving from the framework of how it all fits together, including relationships between the destination and a number of generating countries (i.e. the system) and how the system operates (the process).

Defining and describing tourism

While a systems approach helps us to organise our thoughts about this multi-disciplinary subject, it is not a substitute for a signification of tourism which may be defined and described in many ways. Table 2.1 sets out some ideas from authors who have written about tourism (it does not include tourism anthropologists who have their own list in Chapter 5). I have deliberately avoided the standard formal definitions put out by the World Tourism Organisation; these are easy to find and have been widely published. The purpose of the list is to create an understanding of the complexities of tourism.

There are of course other definitions, involving distance, time, overnight stays, geographical perspectives etc. However, the variety of definitions and descriptions should not necessarily be seen as problematic. As long as the underlying assumptions and purposes that frame the definition are understood then the explanation can

Table 2.1: Descriptions and definitions of tourism

Author	Definition/description	Commentary
McKean (1977)	A profound, widely shared human desire to know 'others' with the reciprocal possibility that we may come to know ourselves . . . a quest or an odyssey to see, and perhaps to understand, the whole inhabited earth	tourism as a positive act of self-fulfilment
Jafari (1977)	A study of man away from his usual habitat, of the industry which responds to his needs, and the impact that both he and the industry have on the host socio-cultural, economic and physical environments	use of word 'man' dated, but shouldn't distract from useful holistic nature of the definition
Mathieson and Wall (1982)	Multi-faceted phenomenon which involves movement to and stay in destinations outside the normal place of residence; comprises dynamic, static, and consequential elements	useful overview, but too broad. Is tourism a 'phenomenon'?
Pearce (1982)	Tourism may be defined as the loosely interrelated amalgam of industries which arise from the movement of people, and their stay in various destinations outside their home area . . . Tourism is, in essence, a phenomenon concerned with the leisured society at play	an under-used but effective definition but makes no allowance for impacts
Murphy (1985)	The sum of . . . the travel of non-residents (tourists, including excursionists) to destination area, as long as their sojourn does not become a permanent residence. It is a combination of recreation and business	concentrates on the purpose of travel

Table 2.1: continued

Author	Definition/description	Commentary
Urry (1990)	How and why for short periods people leave their normal place of work and residence. It is about consuming goods and services which are in some sense unnecessary. They are consumed because they supposedly generate pleasurable experiences which are different from everyday life	focus on consumption as part of the postmodern experience, thus unintentional Eurocentrism?
Ryan (1991)	Essentially, tourism is about experience of place. The tourism 'product' is not the tourist destination, but it is about experience of that place and what happens there: [which is] a series of internal and external interactions	humanistic and experiential allowing for both 'host' and 'guest'
Leiper (1995)	Tourism comprises the ideas and opinions people hold which shape their decisions about going on trips, and where to go . . . and what to do or not do, about how to relate to other tourists, locals and service personnel. And it is all the behavioural manifestations of those ideas and opinions	definition of tourism bounded by tourist behaviour and interaction with the psychological environment, doesn't allow for the industry that responds to them
Middleton (1998)	Although travel and tourism is invariably identified as an 'industry' it is best understood as a total market . . . [which] reflects the cumulative demand and consumption patterns of visitors for a very wide range of travel-related products.	focus on business and the tourist as 'customer'. Fails to recognise impacts. This is tourism as promoted by the WTTC

be taken in its proper context. Thus from the above, we would want to know that Jafar Jafari is interested in the sociology of tourism, that Mathieson and Wall's definition comes from their book about impacts, Urry is a proponent of post-modernism,[1] and Middleton's description is geared towards promoting an industry lobby group whose ideas are fixed on growth: the World Travel and Tourism Council (WTTC).[2]

Characteristics of tourism

As well as definitions and descriptions of tourism, there is also a range of characteristics which have been published. For example, in discussing tourism and post-modernism, Urry (1990:2) may be paraphrased as follows:

> Tourism is a leisure activity which presupposes its opposite, namely regulated and organised work; tourism relationships arise from a movement of people to, and their stay in, various destinations; the journey and stay are to, and in, sites which are outside the normal places of residence and work; a substantial proportion of the population of modern societies engages in such tourist practices; places are chosen to be gazed upon because there is an anticipation, especially through day-dreaming and fantasy of intense pleasures . . . anticipation constructed and sustained through a variety of non-tourist practices, such as film, television, literature, magazines, records and videos, which construct and reinforce the gaze; an array of tourist professionals develop who attempt to reproduce ever-new objects for the *tourist gaze*.[3]

From the above, it can be seen that Urry is interested in analysing the motivation behind touristic travel, not in the way a marketing person might be concerned with motivation (so as to trigger a purchase impulse) but rather in the sense of motivation as a form of social response to the post-modern condition (i.e. the sense of alienation from nature and rejection of that which has gone before to be found in post-industrial societies).

Pearce (1989) has identified other characteristics of tourism. First, as interrelationships are transitory there is a little chance for understanding between hosts and guests. Second, the fact that the tourists are on holiday while the hosts are at work serving them. This is also a point emphasised by both Cohen (1972) and Nash (1977; 1981). They say that this *leisure-service distinction* is universal in relationships between tourists and hosts. Third, there is the element of seasonality which is of importance because it tends to be far more socially disruptive than year-round activities associated with other, non-seasonal economic sectors.[4]

Another characteristic of tourism is its inexorable links with culture. Bryan Turner writing on post-modernism, globalism and Islam, has this to say on cultural tourism:

> Tourist fantasy permits the self to assume diverse social roles in exotic settings; tourism invents and demands empathy to play out short-term fantasy roles. Tourism tends to make cultures into museums, as cultural phenomena which can be viewed as quaint, peculiar and local. Tourism paradoxically is a quest for authentic local cultures, but the tourist industry, by creating the illusion of authenticity, in fact reinforces the experience of social and cultural simulation. The very existence of tourism rules out the possibility of authentic cultural experience.
>
> (1994:185)

This type of critical deconstruction of tourism and tourists is not new, nor is it original. The underlying assumptions about the strength of culture, what constitutes 'authentic' and what tourists do on their vacation are weak and bound up in a very introverted, unconnected view of the world. For example, Louis Turner and John Ash (1975) were saying pretty much the same thing in their polemical book, *The Golden Hordes*. What these views do is to deny the diversity of tourism. This extract from Turner also alerts us to a serious problem in the analysis of tourism by social scientists. Much of their work seems to be bounded by culture and class boundaries. It is useful to ask whether there is a sort of snobbery about what they write concerning 'mass' tourism and their social construct or assumption that it is somehow wrong and not even liked by consumers. This conjecture does not seem to be based on empirical research: they are opinions that arise from working and living in an essentially white, middle class environment.[5] A sense of this comes across in the work of Jacqueline Waldren, an Oxford anthropologist, who has lived in a Mallorcan village for a number of years. She fails to overcome a problem she identifies in the introduction to her book concerning doing 'fieldwork at home' (i.e. separating herself from the people and activities she describes) in her adopted home in the village of Deià). The work has, at times, an unfortunate flavour of Peter Mayle's *A Year in Provence* about it (see for instance Waldren's descriptions on pages 102–3 about hiring and using local tradesmen). Underpinning the book is a sense of snobbery; for a start, the village is no ordinary one. It is an artists' and writers' colony of long standing: Waldren speaks of a guidebook from 1888 which refers to 'its collection of strange and eccentric foreigners' (Waldren, 1996:156). The writer and poet Robert Graves lived there for a period where he 'endeavoured to create a "perfect community": organising social life for artists and writers and their local friends in Deià, and publishing a literary magazine' (ibid).

Just as change in a village is complex, fragmented and multi-faceted, within tourism there is increasingly less a single, stable hierarchy of styles of tourism. Rather a fragmentation of competing styles reflecting age groups, occupational class and lifestyles is emerging as the phenomenon itself matures and marketing expertise becomes more sophisticated. Pfaffenberger (1983) supports this view by looking at changes in motivation in countries moving away from traditional forms of travel such as in Sri Lanka where modern leisure outings are justified as pilgrimages and also in Japan whereby domestic tourism to shrines is accompanied by newer forms of recreation.

While leisure tourism (i.e. holidaymaking) might be thought of as a system for managing pleasure, it is infinitely more complex than this. Tourism has a history of submission and extraction, 'willing' destinations 'submitting' to local élites and multinational corporations, so the debate about tourism and its management must be informed by topics that include:

- the consequences of tourism affecting how heritage and history is told at a particular location;
- the relationships between tourism marketing (especially representation of ethnic[6] cultures by the state), self-identity and an increasingly homogenised 'smaller' world;
- (following on from this) the paradox of tourism's potential to provide the motivation and resources to defend ethnic or group identity ranged against the idea that mass tourism is unlikely to lead to solidarity and understanding between diverse ethnic groups;
- the actual and potential impacts of increased tourism in crowded, multi-communal urban areas where problems and tensions may already exist;
- how to disaggregate tourism from wider influences and socio-economic technological phenomena; (and)
- the potential of tourism to contribute to the development of civil society through the empowerment of equitable economic development.

These discussion topics are unlikely to be resolved through technological solutions. 'Management fixes' rarely take into account the human factor. They tend to see problems in isolation. People cannot be 'fixed' and it is people that create tourism and its problems.

Key ideas

- While tourism can be seen as an industry (or a set of inter-linked industries) it is also a complex set of social phenomena;
- for the purpose of understanding it however, it is best approached as a system (there are a variety of models available) or a set of sub-systems;

- the systems approach forces us into thinking about tourism as being connected with society and cultural process not merely as an economic process; (and)

- if tourism is the 'conversion of dreams' then questions must be asked about what sort of dreams and whether or not the conversions will create, ameliorate or irritate cultural tensions.

Questions

1 Contrast the descriptions of tourism by two authors listed in Table 2.1 (above). For example Pearce (1982) and Middleton (1998).

2 What are the differences (and thus implications) of the systems approach portrayed by i) Poon (Figure 2.4) and ii) Burns and Holden (Figure 2.5)?

3 Describe the tourism system at a destination known to you.

4 To what extent has the intellectual reflections on tourism helped or hindered our understanding of it?

Key readings

There is an alarming array of tourism books that attempt to explain it from various perspectives. Professor Chris Cooper *et al.*'s *Tourism Principles and Practice* provides a very straightforward view of the matter as does the fourth edition of Holloway's *The Business of Tourism*. For alternative views, Burns and Holden's *Tourism: a New Perspective* is recommended. In addition, Polly Pattullo's *Last Resorts* tells the story (in a journalistic style) of tourism in the Caribbean, and Bryan Farrell's *Hawaii: the Legend that Sells* does the same for that particular island state.

There are a number of tourism journals such as *Annals of Tourism Research* (ATR), *Tourism Management*, *Tourism*

Recreation Research, and *Progress in Tourism and Hospitality Research*. These provide the publishing outlet for empirical research and ground-breaking ideas.

3 Tourists

Plate 3: There are many ways of defining tourists and describing their activities. Here, young tourists enjoy a snack of *kai koso* (a traditional Fijian dish) prepared for them during a pre-arranged village trip. The encounter, organised by the Fiji Rucksack Club, was slightly unusual in that the visitors were living in Fiji at the time, and many of the villagers worked in Suva alongside them. Anthropologists have attempted to categorise tourists according to motivation (purpose of visit). This has not always proved successful as such motives can be too complex to explore without deep level one-to-one discussions. Post-modernists claim that one such motivation is the search for pre-industrial life that can no longer be found in the industrialised countries (Paradise lost, perhaps?).

Overview, aims and learning outcomes

The purpose of this chapter is to examine the ways in which tourists have been defined, described and classified. Without stepping into the domains of psychology or marketing, which are the subject of other books and other authors (cf. Pearce, 1982; Seaton and Bennett, 1996), the ways in which motivation has been ascribed to tourists by anthropologists and other academics is discussed. In particular, this chapter aims to:

- investigate various modes of tourist experience;
- develop a preliminary discussion about tourist motivation; (and)
- discuss the notion of tourism as consumerism.

After reading the chapter you should be able to:

- evaluate the different approaches to classifying tourists;
- identify certain dichotomies in interpreting tourist motivation; (and)
- draw the links between tourism, tourists and consumerism.

Introduction

However one defines, describes or analyses tourism, it is the tourist that remains at the heart of the matter. It is the action of a tourist picking up the phone to call the travel agent or getting in a car for a trip that triggers the complex set of servicing mechanisms and impacts that comprise tourism. The key element in this decision to travel that brings together a diverse group of tourism researchers, teachers, marketers and practising professionals is motivation. Anthropologists have particular views about this which are discussed in Chapter 5 (The Anthropology of Tourism) but in the meantime it is useful to set a generalised context about travel motivation and how tourists may be categorised.

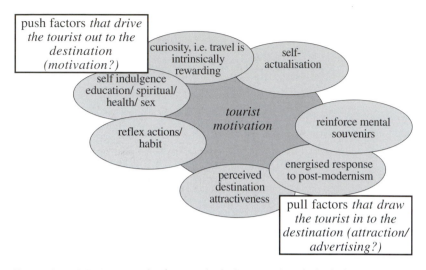

Figure 3.1: Motivation: the forces which direct and underlie behaviour

However, before considering tourist motivation, there is the overarching question about what generally motivates members in society to travel. Figure 3.1 introduces some of the elements.

It can be seen that many of these determinants can be categorised into what marketing specialists call 'push' and 'pull' factors.

Ryan (1991) developed several categories of motivation which he usefully termed the *determinants of travel demand*. These are shown in Figure 3.2:

Definitions and typologies of tourists

As with tourism, the word 'tourists' has a number of formal definitions usually making the distinction between those that travel for a day or less (excursionists) and those that travel overnight or longer. This is the sort of classification made by the World Tourism Organisation and national tourism organisations of national governments. Such a definition is all-encompassing, more or less counting everyone that travels (with the exception of paid workers, migrants etc.), so tourist numbers are always shown as the highest

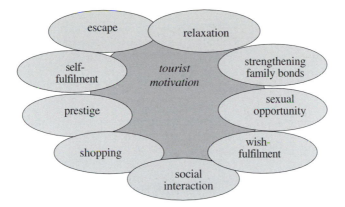

Figure 3.2: Psychological determinants of travel demand
Source: after Ryan, 1991

possible number.[1] The distinction is an important one when planning for tourism at the destination: excursionists will make no demands on accommodation but will use other infrastructure and transport facilities, which they may, or may not, pay for in one form or another.[2] Business travellers may well add value to a hotel's revenue by using laundry or making international phone calls, but may not otherwise engage with the local economy. Likewise, a backpacker may stay in family-owned accommodation and spend money in rural areas not much frequented by other tourists.

However, moving on from a descriptive to a more analytical approach, one of the earliest and most well known is Erik Cohen's tourist typology. This is shown in Table 3.1. He suggested that the key to understanding behaviour and to some extent motivation, is whether or not the tourist is 'institutionalised' (i.e. the extent to which their travel experiences are organised by intermediaries within the tourism industry). He makes the obvious point that different types of such tourists will place different demands on destination locations, some of which will be very intensive.

Cohen's typology provides something of a framework for under-standing destination impacts as well as tourist types. The drifter and the explorer with their longer stay, will have a more profound

Table 3.1: Cohen's tourist typology

organised mass tourist	highly dependent on an 'environmental bubble' created, supplied and maintained by the international tourism industry. Characterised by all-inclusive, fully packaged holidays. Familiarity dominates; novelty non-existent or highly controlled
individual mass tourist	these will use the institutional facilities of the tourism system (scheduled flights, centralised bookings, transfers) to arrange as much as possible before leaving home; perhaps visiting the same sights as mass tourists but going under their own steam
explorer	the key phrase here is 'off the beaten track' perhaps following a destination lead given by a travel article rather than simply choosing from a brochure. This type will move into the bubble of comfort and familiarity if the going gets too tough
drifter	this type of tourist will seek novelty at all costs: even discomfort and danger. They will try to avoid all contact with 'tourists'. Novelty will be their total goal; spending patterns tend to benefit immediate locale rather than large companies

Source: Cohen, 1974

understanding of the hosts and their culture (so Cohen's work says) and the novelty aspect will be dominant.[3] Organised mass tourists will have, so it is said, minimal engagement with the locals and relatively little inter-cultural contact as they remain in the 'environmental bubble' (perhaps we could also call it a psychological bubble) of the familiar. Current thinking on this approach is that it is somewhat deterministic (i.e. pre-determined) in that it ascribes seemingly inevitable characteristics to groups of people. Also, it fails to account for individuals being, over time, several types of tourist as they undertake different travel experiences.[4]

Other typologies of tourists may be found. Stanley Plog (1977) researched on behalf of 16 airline and travel companies to find out if an understanding of travel motivation could help them expand their business. He coined the terms *allocentric* travellers (referring to those who actively seek out the exotic or 'untouched' destinations) and *psychocentric* types (who are not risk-takers and tend to go to well-established tourist destinations). This whole approach comes under the framework of psychographics, where people are categorised according to life-style, self-image, attitudes towards life and social institutions etc.[5]

Also from an industry perspective, American Express commissioned a market study (involving over 6,500 respondents in the US, UK, Germany and Japan) to help them understand some of the underlying thinking of a consumer's travel purchase. The results are not startling, but are shown here as Table 3.2 so as to contrast with the academic approach of Cohen.

In addition to tourist types, classifications by purpose of travel have also been established. Wahab (1975) produced such a scheme based on: recreational tourism; cultural tourism; health tourism; sport tourism; and conference tourism. The advantage of this is that it does retain a focus on what people generally understand to be tourism.

Table 3.2: Amex tourist traveller classification (1989)

adventurer	affluent and educated, these tourists like to try new experiences and meet new people. Travel plays a central role in their lives
worriers	lacking in self-confidence and in their own abilities to travel successfully, this group is nervous about flying and tends to take domestic holidays. They see travel as stress-laden
dreamers	have high aspirations about travel and exotic destinations which are not usually borne out in the actual travel experience which tends to be to rather 'ordinary' destinations. Tend to place great value on maps and travel books
economisers	for this group, travel is not perceived as something that adds particular value to their lives, they engage in it because it is a 'normal' way of taking routine relaxation. They see no worth in paying extra for special amenities or service
indulgers	generally wealthy travellers who will pay for extra comfort and better service. Tend towards staying in five star accommodation, they like to be 'pampered'

Table 3.3: Smith's typology of tourists (1977)

explorers	very limited numbers of travellers who do not see themselves as tourists, and live as active participants and observers among the local people, easily and fully accommodating to and accepting the lifestyles of and norms of their hosts
elite tourists	also a few in number, rarely seen, with individuals who have been 'almost everywhere' but with pre-arranged service facilities and adapting fully but temporarily to local norms
off-beat tourists	uncommon but seen and seeking either to get away from the tourist crowds or heighten the excitement of their vacation by doing something beyond the norm. In general, they adapt well
unusual tourists	occasional in number, travelling in an organised tour and buying an optional one-day package tour to visit some native Indians, generally interested in the 'primitive' culture, but with his 'safe' box lunch, and adapting somewhat to local norms
incipient mass tourists	are a steady flow of people seeking Western amenities and comfort
mass tourists	are a continuous flux of visitors of middle-class income and values, expecting trained multi-lingual hotel and tourist staffs to fulfil their needs as wanted. They obviously expect Western amenities
charter tourists	are groups that arrive *en masse*, who have minimal involvement with the people and culture of the visited country, and who demand Western amenities

Valene Smith, following in the footsteps of Cohen, identified seven types of tourists in a typology (shown here as Table 3.3) which is particularly relevant for social and cultural impact studies.

While each of these typologies has added something to our understanding of tourists, it becomes obvious on analysing them that they add very little to a deeper understanding of tourists.

A group of informed adults could come up with similar typologies based on nothing but individual travel experiences. They all seem riddled with stereotyping which, if based on observations of say ethnic minorities or women, would be seen as being alarmingly racist or sexist. Tourists may choose a destination for many reasons not necessarily just one. They may not even choose a destination as such, but rather a holiday type. For example, young British holiday-makers on a Club 18–30 vacation where social activities (clubbing, drinking, casual sex) with like-minded peers dominate and the destination (provided it has the basic attributes of being relatively near and cheap, having reliable sunshine and reasonable beaches) seems almost irrelevant.

A number of authors have linked tourism with *consumption* (Greenwood, 1989; Burns and Holden, 1995; Ritzer, 1998). Selwyn (1996) draws the inference that as material and spiritual traditions interact on a commercial level with tourists, culture undergoes 'a process of commoditization under the influence of tourists culturally drenched by commodity fetish' (1996:14). MacCannell goes as far as likening the 'all-consuming tourist' with a sort of symbolic cannibal, where tourists consume not only resources and material goods but the very cultures in which they are located, thus paralleling one of the motivations for some types of cannibals: to subsume or incorporate certain characteristics of the victim such as strength or endurance. MacCannell thinks that the modern 'tourist–cannibal' 'manufactures an aura of involvement with the world outside itself while neutralizing it. Cannibalism in the political–economic register is the production of social totalities by the literal *incorporation* of otherness'[6] (1992:66, italics in original).

While MacCannell is describing his personal interpretation of the tourist within a globalising and shrinking world, the links with consumerism are quite clear. Figure 3.3 shows one particular view about how consumers constantly evaluate (though not necessarily at any depth) their responses to products in the market place.

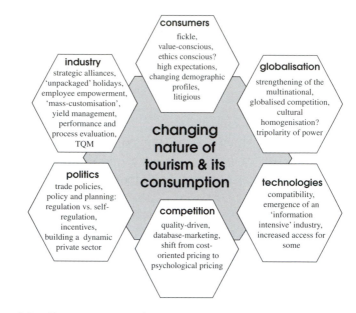

Figure 3.3:   Changing nature of tourism and its consumption

Tourists, mass production and consumerism

One of the biggest cultural shifts seen in countries with advanced economies is the change in attitude towards spending and buying, the change from generally buying what was necessary to the idea of shopping for shopping's sake. Continuing, for a moment, MacCannell's theme from above, 'Consumers in capitalist societies today know that they do not need what they buy' (1992:67). This is evident in the premium prices paid, for example, in sports shops for Nike trainers, the ownership of which can be described as symbolic consumption (in that the price paid reflects neither function nor cost of materials but rather social standing within a restricted peer group).

Bayley relates the history of the word 'consumer' to the development of the Western economy. He describes how the change from 'buyer' to 'consumer' came about:

Mass production and all that it entails, investment, long lead-times, low unit costs and ready availability, replaced a system where simple makers could articulate and satisfy needs; the new distant customers alienated from the production process became consumers.

(1991:47)

One of the most universal terms applied to the phenomenon of contemporary tourism is the expression 'mass'. The movement of peoples across international boundaries has facilitated the development of an industry serving a wide cross-section of the public and meeting the needs of the post-modern consumer.

For tourism development to fulfil the needs of mass consumerism (the mass tourists), many of the characteristics needed for mass consumption referred to by Bayley must be fulfilled. These will include:

- mass production and trans-global repetition of services to meet the needs of hundreds of millions of people moving annually around the globe;
- investment from governments to provide the necessary infrastructure and financial incentives essential to attract corporate investment from global organisations viewed as necessary to provide the facilities for tourism;
- long-lead times by governments who wish to have tourism master plans drawn up and trans-national corporations associated with strategic planning;
- low unit costs achieved through economies of scale as a result of horizontal, vertical and diagonal integration (Poon, 1993) and increasing employee productivity partly through the use of sophisticated communication technology providing a global system of information and reservations for ready access; (and)
- the standardisation of products such as package holidays which may be purchased from the travel agent or by telephone or internet with the minimum of inconvenience.

Thus, the provision of goods and services to tourists and the characteristics of international tourism are framed by mass consumption on the part of nationals from the most developed countries benefiting from low production costs. For example, salaries and wage costs in Food and Beverage departments as a ratio to total revenue in Australia are 32.4 per cent compared to 12.4 per cent in China (PKF, 1993:48). Up to 80 per cent of all international travel (measured by volume) is made by nationals of just 20 countries (WTO, 1998). It is partly this scenario that provides the 'cheap' holiday in Third World countries.

These tourists are also travelling greater distances as economies of scale and technology have pushed the peripheries of travel further outwards from the major source areas of Europe. With the continued advancement of new technology and employment improvements in the generating countries (such as increased holiday leave and disposable income) the demand for long-haul travel has increased substantially. A powerful industry, in terms of its influence on the economies of developing countries, has developed to service the needs of these tourists.

Of particular significance to the tourist is the role of the travel intermediaries (loosely defined as the travel trade) in providing the link between demand and supply. It is these intermediaries who have had a major influence in the development of destinations, influencing speed of growth, type of development and the markets it will serve. As Cooper *et al.* put it:

> The principal role of intermediaries is to bring buyers and sellers together, either to create markets where they previously did not exist, or to make existing markets work more effectively and thereby to expand market size . . . In all industries the task of intermediaries is to transform goods and services from a form which consumers do not want, to a product that they do want.
>
> (1993:189)

The most apparent intermediary is the tour operator who usually puts together, at a most fundamental level, the accommodation, transport, and the ancillary services, into a package which may be bought by the consumer. However, while in the past the tourist had become dependent upon the intermediary (given the difficulties of buying travel products directly from producers) the rapid growth, and ease of access to information technology is changing this traditional position.

Key ideas

• Understanding what motivates tourists to travel (in all senses, not merely the commercial aspect) is at the heart of understanding tourism;

• tourists may be classified in a number of ways (typologies) including the extent to which their trips rely on a series of existing institutions, personality type, and purpose of travel;

• many of these typologies are simplistic, not based on empirical evidence and have become less useful over time in analysing tourist behaviour and motivation; (and)

• there are a number of links which make connections between tourist consumption patterns and so-called 'consumerism'.

Questions

1 Is all behaviour motivated, or is it instinctual? (This is about the nurture versus nature debate.)

2 Why should social scientists be interested in tourist motivation?

3 Are there ways in which 'going on holiday' can be a lifestyle statement?

4   In what ways might mass-tourism have more impact than mass-consumerism?

5   Do tourists conceptualise the world as a sort of single place (the-world-out-there)? If so, is a function of tourism to help the tourist confirm the difference of their own culture?

6   What deeper meaning can be constructed about the seemingly simple phrase, 'I need a holiday!'?

Key readings

Philip Pearce's book, *The Social Psychology of Tourist Behaviour* (1982) is an excellent starting point. It covers the social role of the tourist, approaches to tourist motivation and has a chapter on social contact between tourists and host communities. It is all the better for being more scientific and less judgmental than other books of its era. Georg Simmel's essay on 'The Stranger' in his book, *The Sociology of Georg Simmel* (Wolff, 1950) helps explain the notion of an ambiguous person who exists aside from ordinary society, never joining it. Tom Selwyn's excellent edited volume: *The Tourist Image: Myths and Myth Making in Tourism* (1996) has a noteworthy collection of essays on the theme of tourists and their interaction with the world about them. A number of tourism marketing books necessarily contain chapters about tourist motivation including Seaton and Bennett's *The Marketing of Tourism Products* (1996).

4 Culture

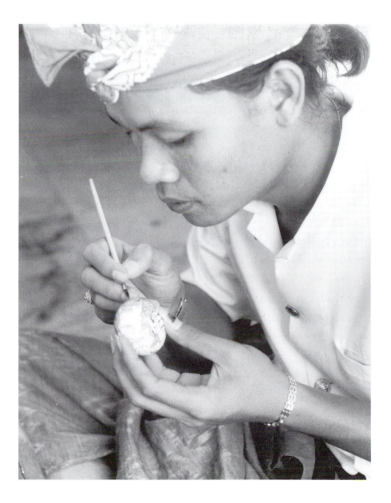

Plate 4: Observing culture represents a major factor in tourism and provides the framework for a constant paradox: does tourism harm culture? The answer is not simple. Here in Bali, an artisan turns from carving wooden objects for sale in a local market to carving fruit for a hotel buffet. The level of skill required and given remains much the same. Could it be argued that in this case tourism has provided an extra outlet for such skills? For anthropologists culture remains a fundamental and central concern. It covers family relationships, ethnic identities, technology, gender positions, migration and exclusion: all that makes up human society.

Overview, aims and learning outcomes

In Chapter 2, it was established that a common thread woven through any tourism system was the idea that understanding something about the nature of culture[1] is an essential factor in gaining an understanding of tourism. In particular, this chapter aims to:

- identify different ways of defining or describing culture;

- evaluate the significance of culture within a tourism system; (and)

- explore some of the links between cultural dynamics and post-modernism.

After reading the chapter you should be able to:

- have an understanding of what culture is and how it evolves;

- describe elements of culture and how these might relate to tourism.

Introduction

At one level, it might be said that 'culture is everything' including socially learned experience, social institutions, science, art etc. The list goes on and of course includes aspects of both receiving and being tourists. However, simple definitions like the foregoing are apt to allow the word or concept they are defining to become taken for granted. Certainly this applies to 'culture'. The danger is that if we stop the discussion at 'culture is everything' then discussions about culture lose their deeper meaning. Factors such as the socio-cultural relations between 'hosts' and 'guests' or where culture can be seen as a potential component for product development, and the ways in which society at a particular destination is structured remain of great significance to those studying tourism.

The concept of culture has a range of meanings according to context. It may mean high art such as to be found in galleries or museums. A person who is familiar with art and music is said, in some societies, to be 'cultured'.[2] This however, is a particular use of the word, it does not define what culture is. Tylor introduced his book *Primitive Culture* (1871) with what has become one of the most quoted definitions, 'Culture or civilization . . . is that complex whole which includes knowledge, belief, art, moral law, custom, and any other capabilities and habits acquired by man as a member of society.'

This remains important because the sub-text it carries is that culture is about far more than material culture. The key word in Tylor's definition is 'acquired' making a very clear distinction between those characteristics that might be biologically inherited and those acquired through learning. Thus in this definition culture is about the interaction of people and how they learn from each other. It promotes the idea that learning can be accumulated, assimilated and passed on through a range of oral and written traditions. The inference to be drawn here is that culture is observed through both social relations and material artefacts. It consists of behavioural patterns, knowledge and values which have been acquired and transmitted through generations.

Geert Hofstede (1991) describes different layers of culture:

- a national level according to one's country . . . ;
- a regional and/or ethnic and/or religious and/or linguistic affiliation [given that nations can be composed of different regions with unique ethnic/language/religious groupings];
- a gender level . . . ;
- a generation level . . . ;
- a social class level, associated with educational opportunities and with a person's occupation or profession; [and]
- for those who are employed, an organizational or corporate level according to the way employees have been socialized by their work organization.

(1991:10)

There is of course a potential problem with Hofstede's first observation about identifying culture with a nation. If we agree that culture is specific to certain groupings or a society, then we must also agree that for a modern pluralistic, complex nation-state (such as Indonesia, Malaysia, the United States or Britain) there is no single culture. This leads into another murky area: the notion of culture and race. In the nineteenth century, ideas about culture and race got very muddled. Cultural characteristics were ascribed to race; biological observations about skin colour etc. were mixed in with cultural observations about how people interact with nature and each other. So while the term 'race' might have a biological meaning such a concept is of little use to the sociologist.[3]

So, while Hofstede's work allows us to interpret some of the relationships within a society, it needs to be contextualised by looking at what are generally seen to be the *components* of culture. These are shown in Figure 4.1.

Given that anthropologists would agree that society (and thus culture) changes in response to environment and technology, the clear implication of this passing on of knowledge and behaviour

Figure 4.1: Components of culture

through generations is that culture is dynamic: all cultures change over time. This is particularly important to remember when attempts are made by tourism planners or tourism companies (or even misguided aid agencies) to 'preserve' culture.

Tourism and culture

The complexity of tourism's social and economic dynamic, both as an act and as an impact, means that it cannot be seen as an integrated, harmonious and cohesive 'whole'. The intention of this section is to develop the systems approaches to tourism (as argued in Chapter 2) by focusing on the particular relationship between culture and tourism: the cultural dynamics, systems and structures that make meaning between visitors and the visited possible. This approach is important in two ways:

- culture can be seen as a commercial resource, especially culture that is perceived to be unique or unusual by actors including tourism marketing specialists and planners; (and)
- understanding the links between tourism systems and culture might help prevent or minimise negative impacts on a host culture occurring through the act of receiving tourists.

There are of course major difficulties associated with trying to think separately about generalised impacts on people caused by the broader process of 'modernisation' from the specific impact of tourism and tourists upon culture. Figure 4.2 shows some of the pressures on any relationship between tourism and culture.

If tourism is seen as a sort of aggregate of business and infrastructure elements (as described in some of the business-oriented approaches to tourism systems such as Poon, 1993) then it becomes impossible to tease out any meaning to the cultural impacts of tourism. This is where the work of Robert Wood shows its importance. His perspective on the issue of culture, tourism and impacts relies on identifying and understanding systems. He states that:

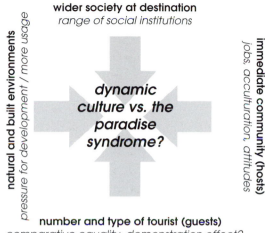

wider society at destination
range of social institutions

natural and built environments
pressure for development / more usage

immediate community (hosts)
jobs, acculturation, attitudes

dynamic culture vs. the paradise syndrome?

number and type of tourist (guests)
comparative equality, demonstration effect?

Figure 4.2: Tourism and culture

the central questions to be asked are about process, and about the complex ways tourism enters and becomes part of an already on-going process of symbolic meaning and appropriation.

(Wood, 1993:66)

'Symbolic meaning' in this sense is a reference to how objects and actions within any society can be given deeper meaning. To use an example from urban society in London in the late 1990s, the popularity of four-wheel-drive vehicles (Range Rovers, Jeep Cherokees, Toyota Land Cruisers etc.) cannot be explained away by simple mechanical need. For the most part, these vehicles never leave the urban/suburban landscape. So ownership and use of these vehicles carry deeper meanings. Perhaps they are a statement about combating 'the urban jungle' – 'it's dangerous out there, I could be mugged, but I protect myself and family through the use of this semi-military vehicle' – or it may simply be a statement of wealth and a demonstration to society at large that you can afford to purchase and run such a vehicle (a form of conspicuous consumption).[4] Coca-Cola was seen by some as a symbol of freedom for many

Eastern-Bloc countries before the USSR and its satellite countries imploded. What is important here is not the 'truth' of such statements, but the fact that they clearly indicate that we are surrounded by symbolic meanings.

Greenwood (1989) offers a more specific sense of the problem between culture and tourism:

> Logically, anything that is for sale must have been produced by combining the factors of production (land, labor, or capital). This offers no problem when the subject is razor blades, transistor radios, or hotel accommodations. It is not so clear when the buyers are attracted to a place by some feature of local culture, such as . . . an exotic festival.
>
> (1989:172)

Given that culture includes place, space and people, then deeper analysis becomes essential so that Greenwood's central concern, the *commoditisation*[5] of culture for tourism, can be addressed. The extent to which components of culture are adapted and offered to tourists for consumption is likely to be framed by at least two factors:

- the relative difference and thus the relative novelty between cultural components of the visitors and the visited; (and)
- by the type and number of visitors.

These two factors help make the link between typologies of tourist types described in Chapter 3 and the broader picture as defined by explanations of tourism systems.

It is culture that distinguishes one group of people from another. In this sense, the culture of the industrialised, Western world can be distinguished from other cultures by a way of interpreting the world about us known as *post-modernism* and a response to this known as *consumerism*. These phenomena are making a vigorous contribution to discussions about tourism, tourists, and motivation and so it is worth briefly exploring the issues.

Post-modernism and consumerism

The dramatic social changes that have been undertaken in the name of rationalisation, efficiency, privatisation and down-sizing usually seen as a part of Thatcherism, has meant (for some parts of the world) the fragmentation of 'culture as a whole way of life' (During, 1993:4). By this, During means a shift from the locally organised and easily understood, to 'culture as organised from afar, both by the state through its education system, and by . . . the "culture industry"' (During, 1993:4). It is not difficult to ascribe international tourism to this 'culture organised from afar' notion and of course tourism is easily recognised as part of a wider 'culture industry'.

These 'organisers from afar' will include powerful producers of cultural forms such as franchised fast foods and sugar beverages that have led to, for example, Ritzer's essay on 'The McDonald-ization of Society' which asserts that bureaucratic structures, previously providing the model for rationalisation, essentially Weber's 'theory of rationalisation' (Ritzer, 1993:xiii) have been replaced by an even more efficient 'rational system' (the fast food chain) with McDonald's being its most important and powerful manifestation. For Ritzer, McDonald's and its clones have become the new social institutions that were formerly thought to consist of family, church, government and the like, their 'rational' approach to production, as Ritzer asserts, has impacted upon hospitals and theme parks alike.

This fragmentation of society can be seen in many ways, but perhaps the rise of post-modernism is the most powerful form (though not all observers of culture would agree with this). During (1993) reckons that there are three grounds for claiming that we live in a post-modern era:

- modernity can no longer be legitimised by the Enlighten-ment ideas of progress and rationality 'because they take no account of cultural differences'
- 'there is no confidence that 'high' or *avante-garde* art and culture has more value than 'low' or popular culture'

- the power of the electronic, global media has pushed distinctions between 'real' versus fake' and 'natural' versus 'unnatural' beyond recognition.

(During, 1993:170)

Here it is useful to think of the repeated images of Marilyn Monroe and cans of Campbell's soup by the American pop-artist Andy Warhol.

The discussion on post-modernism above can be enlarged enabling us to think about the two linked phenomena that underpin and arise from post-modernism: consumerism and commoditisation.

However, I wish to make it clear that for vast parts of the world, including many that are only just engaging in the global tourism nexus, consumerism and commoditisation have no relevance for day to day existence which is dominated by survival and coping. Even so, given that most tourists come from the richer countries of the West (or North), both consumerism and commoditisation collides with the lives of destination residents.

Consumerism impacts on destinations through tourists bringing with them the urban attitudes of the consumer society in which they live. This will include expectations of service levels, and that 'things have their price'. The post-modern controversy lies at the intersection of contemporary cultural change and the political economy of commodity exchange (Shields, 1992:2). Shields, in his introduction to '*Lifestyle Shopping: the Subject of Consumption*' discusses shopping malls as being a post-modern icon and consumerism as *the* post-modern activity. He describes the physical and metaphysical attributes of malls in the following terms:

> The shopping mall . . . is based on one or more levels between two major stores which 'anchor' the dumb-bell shaped plan by providing functional poles of attraction for shoppers. A large food store at one end attracts shoppers from a large department store at the other end . . . The process ensures a steady flow of shoppers . . . strollers, window shoppers and [young and old] 'hangers-out' . . . In the malls, the plan

[is] complex . . . everything is larger, the architecture more monumental (expensive finishes such as marble, skylit arcades, soaring ceiling heights . . .), the major 'anchor stores' multiply and the functions increase with the addition of cinemas, hotels, zoos, recreation complexes . . . in short almost any urban activity one can imagine. Malls now form the architectural typology for office buildings whose elevator lobbies grew first into atria then into malls . . . More insidiously, their 'social logic' of retail capital mixed with the social ferment of crowds of people from . . . all strata forms the model for conceptions of community.

(1992:4)

I am struck that it is possible to assign the characteristics and development ethos of the large integrated malls, to certain Third World holiday destinations and even the new generation of cruise ships, all developed with consumerism (rather than the consumer) at the centre. The notion of 'anchor stores' becomes 'flagship hotel', the 'functional poles' could be golf course and marina. The architecture of modern resorts, though not necessarily high rise, reflects the eclectic 'flashiness' of post-modern mall architecture. The feeling aimed for by developers and engendered by holiday makers at integrated resort developments is reminiscent of Shields' 'conception of community' included in which (in the case of the mall) would be both the *social* function of shopping (as opposed to purchasing) and in the case of the resort or destination, the *social* function of vacationing (meaning the external interaction with other holiday makers as opposed to the internal *re*-creational aspects of vacationing.

There is also the notion that just as shopping malls are 'managed' and the social swirl within them manipulated by design techniques, so are integrated resorts or planned destinations. Management is in the frame, but cannot totally manage the crowd dynamics in the way that they were managed (in the Fordist[6] sense) in Britain's post-war Butlin's holiday camps.[7]

In the above, it can be seen that places of consumption are far more than accidental mixes of geographic locations and attractions. Shields' basic thesis is that they represent both post-modern social dynamics and symbolic edifices wherein everything (including recreation) is for sale: a commodity.

Returning for a moment to Chapter 3, Selwyn reflects on Denis O'Rouke's film about a group of wealthy European and American tourists travelling down the Sepik river in Papua New Guinea[8] and the implications that arise out of the commercial contact between locals and tourists, framed, so to speak, by what he calls a *commodity fetish*. Drawing on a wide range of references including Baudrillard, Eco, and Greenwood, he identifies five assumptions that underpin the commoditisation debate. This is what he says:

1 commoditization is part of a general consumer culture which is itself defined by a culture of unfettered . . . individualism;
2 the imperative for this culture derives from the nature of advanced economic systems which produce unlimited quantities of consumer goods;
3 post-modern consumers resemble either infants [sucking on the mammary gland of consumerism] or schizophrenics [disorganised, bizarre and delusionary behaviour];
4 that cultures defined by commoditization and consumerism are in some specific senses democratic;
5 the commoditization of social and ritual events leads to an erosion of their meaning and that this loss of meaning is accompanied by a parallel loss of feelings of social solidarity; (and)
6 tourist-induced commoditization and consumerism lead inexorably to states of dependency, including cultural dependency, in tourist-receiving regions.

(1996:14)

This may be the received wisdom, but Selwyn goes on to partly demolish these assumptions by asserting that they are too closely linked with redundant ideas about what culture is, specifically that the assumptions are underpinned by the notion that 'exotic' or pre-modern culture shouldn't change: the 'museumisation' approach to Other.

Key ideas

- An understanding of cultures and how they work is of increasing importance for everybody, in all walks of life, as the unseen power of globalisation forces a growing homogenisation of the world;

- it is no coincidence that the word 'culture' has, in its Latin roots (think of 'cultivation'), a connection with the land;

- think broadly about culture so as to include a variety of social behaviours and patterns such as eating, 'personal-space', making love, hair length etc., culture is the glue that holds society together;

- 'nations' do not necessarily equate to 'societies'. Culture and society is not the same as race;

- tourism is one item amongst many that will interact with culture at the receiving destination. There is no such thing as a 'pristine' 'untouched' culture;

- post-modernism is a useful analytical concept for thinking about post-industrial societies, but is probably too Eurocentric (even Narcissistic?) to represent a major leap forward in theorising society as a whole; (and)

- consumerism is a by-product of a broader movement, commoditisation where social institutions and cultural artefacts are seen as expendable in the name of global capitalism.

Questions

1   In what ways is culture used in tourism?

2   What are the connections between culture, consumerism and tourism?

3   Discuss the pressures on the relationship between tourism and culture.

4   Discuss the significance of the fashion label (Levi 501s, Nike trainers, Moschino belts etc.) to youth culture in some societies.

5   What are the links between post-modernism and tourist motivation?

Key readings

Business Studies students might find Geert Hofstede's book *Cultures and Organizations: Software of the Mind* (1991) useful in that it draws links between broad national traits and corporate culture (he mentions tourism briefly). George Ritzer's *The McDonaldization Thesis* (1998) offers fascinating insights into the social phenomenon of consumerism. In this new book there is a short section on tourism and theme parks. Of course, any ethnographic book will be framed and driven by descriptions and analysis of a specific culture, and Jeremy Boissevain's edited collection *Coping with Tourists: European Reactions to Mass Tourism* (1996) is full of cultural insights.

Part II

Anthropology of tourism, globalisation and development

Part I examined separately anthropology, tourism, tourists and culture. This presented a range of factors that generally set them into their broader, and sometimes contentious contexts. Chapter 5 brings them together in the form of an exploration of the possible themes in an anthropology of tourism, notably tourism-as-pilgrimage and tourism-as-imperialism. Leading out of this, Chapter 6 revisits the themes, teasing out of them certain issues that continue to challenge those who study tourism from an anthropological perspective, seeking answers to four basic questions: Is tourism a modern form of religion or pilgrimage? Does tourism damage culture? Can tourism offer Paradise on earth? And, Will tourism bring development? Chapters 7 and 8 take a necessarily different tack with discussions on *globalisation* (picking up on the anthropological theme of local–global relationships) and *development* (given that a major concern for anthropologists is the characteristics of modernisation and development).

5   The anthropology of tourism

Plate 5: This structure seen on the roadside between Habarrana and Polonnaruwa (central–east Sri Lanka) has been built by enterprising villagers as a sort of tourist trap. Its sole function is to provide an arresting sight for tourists to stop and look at. Families then come out from their homes and ask for money in exchange for the right to take photos of it. One villager did offer an alternative explanation saying that it was a safe place for families to run to when wild elephants were rampaging. There was no way for me to verify this, but the laughs on the other family-members' faces seemed to indicate that this was unlikely. Field work by an anthropologist could uncover the 'secret' of this strange contraption and find out how it came about and who gets the benefit.

Overview, aims and learning outcomes

Having examined separately anthropology, tourism, tourists, and culture, this chapter brings them together and discusses how an anthropology of tourism has emerged and who the key authors and academics in this field are. The chapter aims to:

- describe the development of the anthropology of tourism;

- identify the key scholars who have contributed to the anthropology of tourism; (and)

- differentiate the main approaches to the anthropology of tourism.

After reading the chapter you should be able to:

- demonstrate an understanding what the anthropology of tourism is;

- describe tourism from various anthropological perspectives such as rites of passage, cross cultural encounters, cultural inversion; (and)

- be aware of the various scholarly contributions to the anthropology of tourism.

Introduction

Anthropology and tourism (as a field of knowledge) have obvious synergy. Both seek to identify and make sense of culture and human dynamics. Because tourism is a global set of activities crossing many cultures, there is a need for a deeper understanding of the consequences of the interaction between generating and receiving tourism societies (Burns and Holden, 1995). We started this process in Chapter 2 with the examination of various tourism systems. In taking this analysis further, Smith indicated:

Anthropology has important contributions to offer to the study of tourism, especially through . . . basic ethnography . . . as well as the acculturation model and the awareness that tourism is only one element in culture change.

(1981:475)

One interpretation of this may be that Smith is arguing that human interaction, not business and marketing, is the key factor in tourism's many paradoxes. If we accept this, the link between anthropology and tourism becomes all the more important. Taking the key themes that underpin anthropology, it can be seen that it is able to bring to tourism:

- its characteristic *comparative framework* (studying a variety of phenomena in different locations in order to identify common trends);
- a *holistic approach* (taking account of social, environmental and economic factors and the links between all three); (and)
- pursuit of *deeper level analysis* (i.e. what *causes* tourism).

While these themes may seem obvious ones to those already associated with the study and analysis of tourism, the path to establish an anthropology of tourism has not been an entirely smooth one. For example, in an early essay on the subject, Nash (1981)[1] made several observations about the credibility of tourism as a 'serious' subject for anthropological study. These observations have been put together in a schema (Figure 5.1).

Nash (1981) also made the point that it is the cross cultural encounters and the consequential social transactions 'that provide the key to the anthropologic understanding of tourism'. He goes on to explain that this encounter will have many variations, not least of which is that one group (the tourists) are at play, while another group (tourism employees) are at work. We could also add a further group, the local residents, who might be classified as both active and passive observers. Having set something of a framework, the next section deals with the beginnings of the anthropology of tourism.

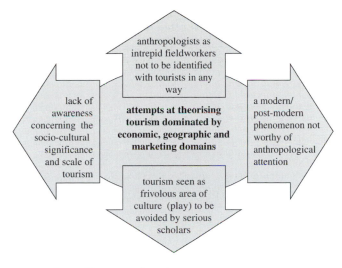

Figure 5.1: Anthropologists in denial!
Source: after Nash, 1981

The roots of the anthropology of tourism: sacred journey or imperialistic exploitation?

Prior to the development of a recognisable anthropology of tourism, elements that now form part of it were being explored by sociologists and anthropologists in a number of contexts. This is illustrated in Figure 5.2.

From these early social scientists, Durkheim (1858–1917), was arguably the most influential in having an eventual impact on the analysis of tourism followed, perhaps, by Van Gennep and Victor Turner.

Durkheim is generally recognised as the scholar who conceived and constructed the theoretical framework within which sociology was able to operate as a science. He considered questions about the relationship between the individual and society. In his essay on *The Elementary Forms of Religious Life* (1915) Durkheim refers to 'present day . . . society constantly creating sacred things out of ordinary ones' and of 'substituting for the real world another different one' and 'systematic idealization' (Abraham, 1973:179;

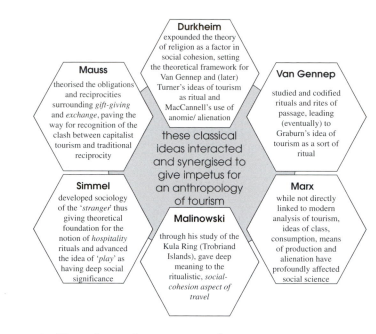

Mauss

theorised the obligations and reciprocities surrounding *gift-giving* and *exchange*, paving the way for recognition of the clash between capitalist tourism and traditional reciprocity

Durkheim

expounded the theory of religion as a factor in social cohesion, setting the theoretical framework for Van Gennep and (later) Turner's ideas of tourism as ritual and MacCannell's use of anomie/ alienation

Van Gennep

studied and codified rituals and rites of passage, leading (eventually) to Graburn's idea of tourism as a sort of ritual

these classical ideas interacted and synergised to give impetus for an anthropology of tourism

Simmel

developed sociology of the '*stranger*' thus giving theoretical foundation for the notion of *hospitality* rituals and advanced the idea of '*play*' as having deep social significance

Malinowski

through his study of the Kula Ring (Trobriand Islands), gave deep meaning to the ritualistic, *social-cohesion* aspect of travel

Marx

while not directly linked to modern analysis of tourism, ideas of class, consumption, means of production and alienation have profoundly affected social science

Figure 5.2:   The early social science roots of tourism

187). Significantly, in trying to 'explain' religion, Durkheim concluded that rituals and religion serve society by producing increased social solidarity. He defined the sacred in the following terms:

> For our definition of the sacred is that it is something added to and above the real: now the ideal answers to this same definition; we cannot explain one without explaining the other. In fact, we have seen that if collective life awakens religious thought on reaching a certain degree of intensity, it is because it brings about a state of effervescence which changes the conditions of psychic activity. Vital energies are over-excited, passions more active, sensations stronger; there are even some which are produced only at this moment. A man does not recognize himself; he feels himself transformed and consequently transforms the environment which surrounds him.
>
> (Durkheim's words cited in Abraham, 1973: 187)

This analysis has been applied to tourism by Nelson Graburn who wrote of 'Durkheim's notions of the sacred – the non-ordinary experience – and the profane' (Graburn, 1989:24).

In similar vein, van Gennep (1960, first published in 1908) theorised about the transition from one social category to another during the so-called life cycle of the individual. His analysis was about the rites of passage that frame singular significant events in an individual's life such as puberty, entry into adulthood, marriage, parenthood, etc. He put forward the idea that such rites of passage consisted of three main elements:

- *separation* (the ritual removal of a person from society and 'ordinary' life as led thus far);
- *liminality* (a period of marginality or seclusion for the person following separation and prior to the next stage); (and)
- *incorporation* (re-aggregation of the person into society with their new status).

It should be noted that there is a distinction to be made between rites of passage which can be defined as going through a one-off process to change social category (typically, and as noted above, from childhood to adulthood, unmarried to married, wife to widow) and rites of intensification which stem from the annual cycle of renewal that re-affirm (or intensify) a person's relationship with society and help mark seasons of the year to change the pace of society on a continuing basis (such as the annual monsoon, birthdays or anniversaries, the harvest season, or the onset of winter).

The rituals that surround a Western style wedding ceremony,[2] the white dress, a stag/hen night out in the company of friends of the same sex, the bride and groom arriving at the church at ritualistically defined times and from different locations, the groom not being allowed to see his bride-to-be (at least on the day of the wedding) until she arrives at the church door etc. are all indicative of van Gennep's three stages.[3] These rituals are said to help with reinforcing collective sentiment and social integration. Victor Turner developed these ideas of transitions:

In the actual situation of ritual, with its social excitement and directly physiological stimuli, such as singing, dancing, alcohol, incense, and bizarre modes of dress, the ritual symbol, we may perhaps say, effects an interchange between its poles of meaning. Norms and values, on the one hand, become saturated with emotion, while the gross and basic emotions become ennobled through contact with social values.

(1967:30)

Thus in some societies, it is, under the special circumstances of a stag-night, 'acceptable' for a man to become uncontrollably drunk and perhaps act in a lewd fashion in a strip-club! Turner refers to these periods between 'normal' spheres of existence (i.e. every-day life) as *liminal* or halfway states and categories. It is relatively easy to see why van Gennep's and Turner's ideas have proved attractive to those searching for the 'why' of tourism. Young Australians and New Zealanders appear to go through a sort of rite of passage when they take their long trip to Europe. It is usually at a stage between university and starting working life. It involves being with peers undertaking a similar trip, is set aside from normal life and usually encompasses some sort of welcome-home 'ceremony' as they arrive back ready to join 'normal' society, follow 'normal' rules which usually includes getting married and 'settling down'.

A key question arising at this stage is that if tourism is a ritual or rite, in the way that van Gennep characterises them, do they have a function in reinforcing collective sentiment and social integration? We can go a step further and ask, in the context of the main tourist generating areas of the industrialised/post-industrialising world, 'what does being a tourist mean to society?'

The answers might be that in seeking to find some special mood of 'sacredness' that demarcates the religious from the profane, tourism could be seen as a totem[4] of freedom (the US Travel and Tourism Administration's strapline for tourism is 'Travel: the Perfect Freedom'). Thereby 'worshipping' tourism as a symbol of modern economic and social freedom, could be interpreted as worshipping society itself[5] – exactly replicating Australian aboriginal use of totems as described by Durkheim in his original field research.

Having examined the idea of tourism as ritual and sacred journey (especially the work of Graburn) we now turn to another perspective: that of tourism as a form of imperialism. Turner and Ash, in their polemic mood, indicated:

> Modern tourism is a form of cultural imperialism, an unending pursuit of fun, sun and sex by the golden hordes of pleasure seekers who are damaging local cultures and polluting the world in their quest . . . Tourism is an invasion outwards from the highly developed metropolitan centres into the 'uncivilised' peripheries. It destroys uncomprehendingly and unintentionally, since one cannot impute malice to millions of people or even to thousands of businessmen and entrepreneurs.
>
> (1975:129)

Nash's work in this area was framed by a similar belief: that the modern tourist, as he puts it, 'like the trader, the employer, the conqueror, the governor, the educator, or the missionary, is seen as the agent of contact between cultures and, directly or indirectly, the cause of change particularly in the less developed regions of the world' (1989:37).

Several themes arise of Nash's essay which have had a powerful influence on the work of many who study tourism from a social science perspective. They are summarised here:

- analysis of touristic development should not take place without reference to the productive centres that generate i) sufficient surplus to enable tourism (in the sense of leisure travel) and ii) tourists themselves;
- this situation sets up a degree of control by the generating region over the receiving region and that this 'makes a metropolitan center imperialistic and tourism a form of imperialism';[6]
- this relationship is totally geared towards supplying whatever

the tourists want including that which may not 'naturally' or 'traditionally' be found such as fast food, air conditioning, swimming pools and imported food and beverages: a supporting infrastructure is thus developed;

• transactions with local people are inherently unequal and that it is this inequality that frames the relationship between 'hosts' and 'guests';

• there are often economic disparities between 'hosts' and 'guests' which, like colonialism, can engender feelings of superiority among the incomers. 'People who treat others as objects are less likely to be controlled by the constraints of personal involvement and will feel freer to act in terms of their own self-interest'; (and)

• a tourism system may develop (especially in countries with a very limited economic base) which may subsume the general economy into a service economy geared towards meeting the needs of transient, leisured strangers and their sponsors.

These themes may be summarised as *dependency* (which is dealt with in Chapter 8) and *acculturation* (as discussed in Chapter 6). However, evidence of empirical research to back up these speculative musings are not particularly strong (see also comments later in this chapter to do with Boorstin's work). Amazingly, Nash calls native people (his term) who take the initiative in helping to develop touristic areas, 'collaborators'. The selection of potential touristic locations were selected, according to Nash:

> with the collaboration of their inhabitants, [and] developed because of their compatibility [i.e. constant sunshine or ski conditions etc.] with metropolitan dreams. Their fate thus became linked with exogenous forces over which they were to have less and less control.
>
> (1989:42–3)

Nash's later work is much less critical of tourism as a social activity, and in his most recent book, published in 1996, the word 'imperialism' does not appear in the index. What Nash and his followers failed to address in their analysis is the question of economic alternatives. Having seen the conditions under which, for example, young women are employed in the clothing factories in the Export Development Zones around Colombo in Sri Lanka, where their every movement is controlled and regulated and wages paid are barely supportable, clearly the garment industry doesn't offer much hope. In this context, tourism provides a far more realistic potential.

Key themes in the anthropology of tourism

In Chapter 1, five key themes in anthropology were suggested (cf. Figure 1.4): the nature of culture; culture and survival; the formation of groups; the search for order; (and) change and the future. The way in which these broad themes interface with tourism can be explained through the current work of Jafar Jafari (editor-in-chief of *Annals of Tourism Research*) shown as Figure 5.3.

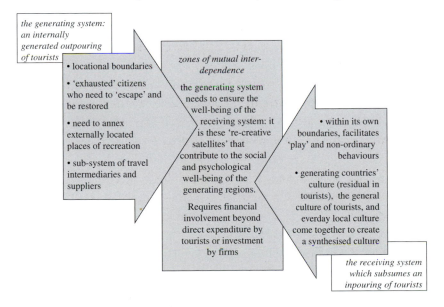

the generating system: an internally generated outpouring of tourists

• locational boundaries

• 'exhausted' citizens who need to 'escape' and be restored

• need to annex externally located places of recreation

• sub-system of travel intermediaries and suppliers

zones of mutual interdependence

the generating system needs to ensure the well-being of the receiving system: it is these 're-creative satellites' that contribute to the social and psychological well-being of the generating regions.

Requires financial involvement beyond direct expenditure by tourists or investment by firms

• within its own boundaries, facilitates 'play' and non-ordinary behaviours

• generating countries' culture (residual in tourists), the general culture of tourists, and everday local culture come together to create a synthesised culture

the receiving system which subsumes an inpouring of tourists

Figure 5.3: An anthropological view of a tourism system
Source: after Jafari (in Witt and Moutinho), 1995

What Jafari is suggesting here in Figure 5.3 is of great interest to social scientists studying tourism. His way of looking at a tourism system is first of all to set up two sub-systems, one for the *generating regions* which provides an 'outpouring' of tourists, and another for the *receiving areas* which subsumes an 'inpouring' of tourists. He then places these two sub-systems within the context of a third: *the zones of mutual inter-dependence.* Jafari makes a connection between supply and demand that goes far beyond economics or marketing. His point is that the industrialised (or post-industrialised) countries depend on the *re-creative satellite areas* to help regenerate exhausted citizens. This is quite a different position from political scientists who will stress the economic failure of tourism to 'deliver the economic goods' to developing countries. However, if existing destinations fail to 'deliver the recreational goods' what remains is the economic power of the generating countries to shift their attention and business to other locations. In Jafari's terminology, they would create more *re-creative satellites.* The next section elaborates the work of some influential writers on the anthropology of tourism.

Key writers in the anthropology of tourism

The above description on the development of an anthropology of tourism and Jafari's insightful perspective has set the scene for a more detailed discussion about the work of a number of academics who contribute to the anthropology of tourism. It is also helpful because it allows us to reflect on why anthropologists should study tourism. Using Nash's arguments, this is spelt out in Figure 5.4.

Given its relatively short time span, it is quite easy to identify a number of important and influential contributors to the anthropology of tourism.

I have not included Jafar Jafari and Valene Smith in Table 5.1 because, in a sense, their work in the anthropology of tourism goes beyond a contribution to theorising about the subject. Between them, and in their own different ways, they have single-mindedly placed tourism on the academic map as a legitimate subject to

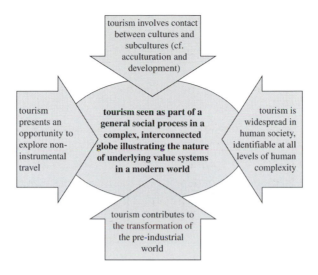

Figure 5.4:   Why anthropologists should study tourism
Source: after Nash, 1981

study. Jafari through his founding and continuing editorship of *Annals of Tourism Research* and Valene Smith through her work with the American Anthropological Association. Drawing up a list necessarily means exclusion: it should be borne in mind that the purpose of the list is to stimulate thinking about the anthropology of tourism amongst students, not to provide the definitive register. For instance, it might be argued that other significant players are Linda Richter (1989) and her work on the politics of tourism, Philip Pearce (1982) with his contribution to the social psychology of tourism; and Roland Barthes' (1984) essay on the Eiffel Tower in which he invites the casual observer to interpret the Paris landscape below (the Bourse, the Seine, Sacré Couer, the Royal Palaces etc.) as part of a mythology that makes a city whole. The list goes on!

The different approaches

If we link Figure 1.4, key themes in cultural anthropology, to Figures 5.3, 5.4 and to some extent to Table 5.1, four broad themes

Table 5.1: Key authors in the anthropology of tourism

Author*	Definition/description/stated position	Commentary
Graburn (1977)	a special form of play involving travel, or getting away from 'it all' (work and home), affording relaxation from tensions, and for some, the opportunity to temporarily become a nonentity, removed from a ringing telephone	tourism as a form of *escapism* or pleasure-seeking
Nash (1981)	resulting from the intersection of the histories of two or more cultures and subcultures . . . It becomes a process involving the generation of tourists, their travel and their subsequent encounter with people in some host society. Such an encounter implies transactions between tourists, their agents and hosts which affect the people and the cultures involved	considers consequential aspects of relationship between supply and demand. Concludes his thoughtful definition with the simple idea that tourism is 'leisure activity requiring travel'
Selwyn (1994)	asks that the anthropology moves from its unsustainable generalisations about 'commoditisation' and 'authenticity' to be more rigorously ethnographic and more theoretical; sees tourism as sets of relationships in the widest imaginable economic, political, social and cultural contexts	this allows for displacement, interpretation of history, authenticity of memory and the political economy of culture to be examined as part of the wider tourism debate
Urry (1990)	links tourism as a cultural practice with post-modernism and the relationship between those that serve and those (mainly middle class) that consume goods which in some senses are 'unnecessary'. Tourism as a contrast with everyday life	allows for a systematic study of tourist motivation from a social science perspective but fails to address the paradox of the many tourists who deliberately seek out the same (familiar food, peers and surroundings) while on holiday

Table 5.1: continued

Author*	Definition/description/stated position	Commentary
MacCannell (1992)	'tourism is a primary ground for the production of new cultural forms on a global base. In the name of tourism, capital and modernised peoples have been deployed to the most remote regions of the world, farther than any army was sent . . . In short, tourism is not just an aggregate of merely commercial activities; it is also an ideological framing of history, nature, and tradition; a framing that has the power to reshape culture and nature to its own needs' (1992:1).	while MacCannell has added much to the theorisation of tourism, it is difficult to believe his central thesis that all tourists are searching for authentic experiences denied to them at home in their industrial/post-industrial world
Boissevain (1996)	'How then do individuals and communities dependent on the presence of tourists cope with the commoditisation of their culture and the constant attention of outsider?' (1996:1). Proposes a series of resident behaviours or 'coping strategies'.	his ideas and generalisations are based on long-term study of how tourism has changed Malta over a 35 year period:
Cohen (1988)	most significant contribution has been the development of the first typology of tourists which signified the need to differentiate between a number of tourist types concluding that 'intellectuals and more alienated individuals will engage in a more serious quest for authenticity than most rank-and-file members of society' (1988: 376).	this aspect of his work is based around the reality of tourist experiences; like other typologies of various types, it has been criticised for being somewhat deterministic
Dann (1997)	the 'language' of tourism (including its imagery); 'tourism research without theory is quite dead'. 'Working separately, theoretitions and practitioners become victims of their own [separate] monologue[s]'.	proposes that theory and academic research must feed into the tourism industry as a practical benefit to contribute to sustainability

Note: *Not all of these writers are anthropologists. Sociologists, such as MacCannell, Dann and Urry have also made significant contributions and are thus included.

emerge as being of prime importance in the anthropological study of tourism: the paradox of being local in a global world; tourism and ritual; tourism as mythological adventure (and) tourism and social change. There are of course other themes (as Selwyn, 1996, reminds us) such as those proposed by Malcolm Crick's categorisation of three strands of enquiry that inform the anthropology of tourism:

- *semiology*,[7] the study of meaning and relationships between an image or symbol (the signifier) and the concept associated with it (the signified) which is formed by a society's denotation and connotation of the particular image (Barthes' essay on the Eiffel Tower, as mentioned above, is a classic example). Tourists occupy cultural space during their visit, and as part of their search for the authentic, place special significance on things, 'marking' them out to be special even if these places or sights might, in themselves, be unremarkable;[8]
- *political economy*, issues of power and control and the forces that shape touristic development at a particular destination. It brings together economic and political domains so as to enable a deeper understanding of the political implications of development and economics; (and)
- *social and cultural change*, within both the generating and receiving areas, in particular commoditisation of place and culture, cultural outcomes of being visited, and tourism as a search for authenticity.[9]

This search for authenticity can be explained as a sort of compensatory process by which 'the alienated worker seeks a less alienated, more authentic existence during a vacation abroad' (Nash, 1996:66). This view is based on MacCannell's work which, in turn, was based on a rejection of Boorstin's thesis (1964) that the modern tourist intentionally seeks out inauthentic experiences, the so-called *pseudo-events*[10] as part of a generally superficial lifestyle at home. MacCannell goes on to assert that for Boorstin, 'there is something about the tourist setting itself that is not intellectually satisfying'. (MacCannell, 1976:103). MacCannell's position on Boorstin's pseudo-events is that tourists do not cause pseudo-events, as Boorstin insists (i.e. that tourism has become superficial because tourists themselves are superficial), but that Boorstin's snobbish attitudes on the distinction between tourists and travellers are 'part of the problem of mass tourism, not an analytical reflection on it' (MacCannell, 1976:104).

Urry (1990:11) broadens this debate about the deeper motivations of tourists by referring to 'general notions of liminality and inversion'. Liminality, in this sense, is meant in the way that van Gennep intended, as a 'release from routinized social structure' (Nash, 1996:41). However, Urry has a view which differs from both Boorstin and Nash. He *rejects* the idea of the search for authenticity as the *key* motivating factor for tourists (although acknowledging that it may be important). He continues with a further comment about the basis for the organisation of tourism, 'one key feature would seem to be that there is a difference between one's normal place of residence/work and [the tourism experience] . . . because there is in some sense a contrast with everyday experiences' (1990:11).

Here again the issue of liminality is being referred to. This uncertainty about tourist motivation emphasises that which was established in Chapter 3: there is no one single type of tourist. The experiences they seek will be different, Nash, in discussing Cohen's analysis of tourists, puts it this way 'these people may seek more or less authenticity in their tourism according to how alienated they are from the social conditions in which they live' (1996:66). He continues in similar vein: 'There are, indeed, other types of tourists (and, one might add, hosts) for whom the issue of authenticity does not come up' (Nash, 1996:82). On this, Urry cites Feifer who assures us that the tourist[11] is not, 'a time-traveller when he goes somewhere historic; not an instant *noble savage*[12] when he stays on a tropical beach . . . Resolutely 'realistic', he cannot evade his condition as outsider' (1990:100–1, italics added).

Selwyn ties up these rather loose ends by reminding us that this idea of being an outsider is in no small way connected with the somewhat Eurocentric search, to coin Maslow's phrase,[13] for 'self-actualisation':

> By now it is widely accepted by anthropologists of tourism that much of contemporary tourism is founded on the 'Quest for the Other' . . . pushing as they do in opposite directions, the quest for the 'authentic Other' and the quest for 'authentic Self' constitutes the 'tension which informs all tourism'.
>
> (1996:21)

This idea of tensions reminds us of Lévi-Strauss' notion of opposite states of being and Malinowski's symbolic dualisms (as noted in Chapter 1 of this book). For tourism, these tensions can be explored even further. There remains however, a need to explain why specific tourist modes are attached to particular social groups at a particular historical period, the unanswered questions are why particular behaviours? Why particular groups? The nearest these become to being answered is through the work of Passariello (1983) who studied middle-class Mexican resort tourism. Three interconnected factors were suggested which both offer an explanation for how the patterns found were generated and could help predict further patterns. These are (with questions added by the present author):

1 Discretionary income limits choice of style, distance and length of travel. *But, is simply having enough money sufficient explanation?*
2 Cultural self-confidence, childhood and educational experiences. But, can we connect middle-class wealth with cultural self-confidence? *Or does this assume that the things that comprise a middle-class education (literature, Greek mythology, the Arts etc.) are the only worthwhile components of culture?*
3 Cultural inversions with meanings and rules of ordinary behaviour suspended or turned on their heads. *But can this excuse child prostitution, drunkenness, indolence?*

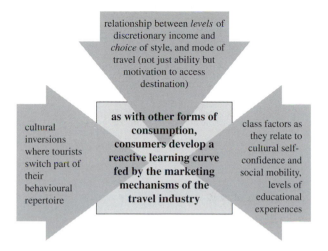

Figure 5.5: Social motivation and tourism consumption
Source: after Passariello, 1983

Passariello's work is important, if somewhat incomplete, because the analysis is based on empirical research rather than reflection. Figure 5.5 shows her ideas in a schema.

In the particular case of Passariello's work, we can see that choice of tourist style stems from their home culture and situation. But, if we generally agree with the notion of inversion and behaviour in tourists, how do people select such changes? Is it simply a choosing of those elements they are not able to change in their normal lives (within constraints of discretionary income and self-confidence)? Given that inversion is a continuing theme in anthropology, it is worth examining it a little more closely as it pertains to tourism. Some of these are shown in Table 5.2. Note that inversions can be in either direction; the polarities are inter-related; tourists usually seek more than one reversal so switch only part of their behavioural repertoire.

It must be stressed that while Passariello's work is useful, it has flaws that have yet to be addressed such as: How can *cultural confidence* be defined? What role does the travel industry have in influencing decisions? (and) Why do some people with 'cultural self-confidence' choose *not* to become tourists?

Table 5.2: Social/cultural inversions

Dimension	Continua
environment	winter–summer isolation–crowds
class/lifestyle	simplicity–affluence thrift–self-indulgence
'civilisation'	nature–urban slow–fast security–risk
formality	nudity–formal clothing sexual restriction–sexual licence
health and tension	tranquillity–stress sloth–exercise ageing–rejuvenation

Key ideas

• Anthropologists and other social scientists argue that *people*, rather than *business* lies at the heart of the need to analyse tourism;

• anthropology offers an approach to the critical analysis of tourism through its comparative framework, the ability to bring the local and global together by recognising the interconnectedness of economic, environmental and social domains;

• there has been a split between those anthropologists who viewed tourism as a ritual (e.g. Nelson Graburn) and those who considered it to be a form of imperialism (e.g. Dennison Nash). The current anthropological thinking is that tourism has many motivations and is too complex to be thus categorised; (however)

• a practical categorisation of the ways in which tourism may be studied by anthropologists has been put forward by Crick who proposed semiology, political economy and social/cultural change as the most effective strands of enquiry;

- too much work on the anthropology of tourism lacks empirical (i.e. research) grounding, and may reflect the white, middle class views of the authors rather than scientific evidence; (and)

- Selwyn and others are convinced that it is the search for Other (an authentic, unspoilt thing 'out there') and the search for authentic self (in the sense of coming to terms with living in a post-modern society) that creates the tension that underpins the social science approach to analysing tourism.

Questions

1   What evidence do we have that specific tourist modes are attracted to particular social groups at a particular time?

2   Why do some people with the means and discretionary income not wish to be tourists? Could it be argued that those who have reached Maslow's 'self-actualisation' state don't need tourism?

3   To what extent is the social scientific analysis of tourism a form of self-indulgence and a form of 'legitimately' laughing at working class consumption of leisure?

4   What part do curiosity, desire for novelty, personality, susceptibility to media play in motivating the urge to travel?

5   Is tourism a form of imperialism?

6   Is tourism a form of religion?

Key readings

The seminal book for the anthropology of tourism has always been Valene Smith's *Hosts and Guests: the Anthropology of Tourism* (1989, 2nd edition). It contains a number of classic essays and has stood the test of time, mainly because it is eminently readable. It has been superseded to some extent by Tom Selwyn's edited volume *The Tourist Image: Myths and Myth Making in Tourism* (1996) which contains such chapters as: 'Genuine Fakes' (David Brown); 'The People of Tourist Brochures' (Graham Dann); 'Passion, Power and Politics in a Palestinian Tourist Market' (Glenn Bowman) and others on the theme of image in tourism. Dennison Nash's book the *Anthropology of Tourism* (1996) while authoritative has proved something of a disappointment in that it doesn't quite put across the passion that can be aroused in this controversial field of knowledge. Dean MacCannell's *Empty Meeting Ground: the Tourist Papers* (1992) has been criticised for its self-indulgent style, but I have found it a great inspiration, especially the essay about Maya Ying Lin's Vietnam Memorial in Washington (the sheer black slab of polished granite engraved with the names of the American military dead, you may have seen it in any number of movies about death and loss in Vietnam). It might be argued that MacCannell's book hasn't got much to do with tourism, but in a sense that shouldn't be a problem: enjoy! Linda Richter's *The Politics of Tourism in Asia* (1989) is a classic text on that subject and given that politics is an integral part of culture, then this book is an important piece of the tourism jigsaw: it is written in a very accessible style. Some of the richest material is (of course) to be found in the journal *Annals of Tourism Research*.

6 Issues in the anthropology of tourism

Plate 6: A controversial advert for the now defunct Club 18–30 holiday company. Based in Britain, this tour operating company made no bones about what it was selling: cheap fun-filled holidays in the sun. The brochures were characterised by pages of hotels that looked pretty much the same. There was no sense of geographic location except when the night club scene was particularly spectacular. Two paradoxes arise out of this, first, it was one of very few brochures to show young black Britons in an integrated setting and having fun. Secondly, it shows the huge gulf that exists between Western youth culture and the rest of society. The emphasis on sex caused many complaints from the public and eventually the advertising campaign was withdrawn. An anthropologist might want to understand more about the role of a young person's first holiday abroad without parents as part of a rite of passage into adulthood. Furthermore, the notion of behaviour reversal (i.e. behaving 'normally' all year and then behaving 'badly' for a two week period away from the constraints of everyday society) would provide a research theme.

Overview, aims and learning outcomes

The purpose of this chapter is to thread together a number of issues within the anthropology of tourism that remain contentious or perhaps unresolved. This will mean revisiting some ideas that have been discussed in previous chapters, especially Chapter 5. The specific aims of this chapter are:

- to create an awareness that tourism can be interpreted as a modern form of pilgrimage;

- to develop an understanding that there is a balance to be struck between seeing tourism as cultural impact and seeing it as just one part of a wider modernisation phenomenon;

- to suggest that tourism carries with it an important social function as a form of modern myth; (and)

- to create a deeper awareness of the local–global relationships forged by tourism.

After reading the chapter you should be able to:

- understand the nature of tourism as 'religion', 'pilgrimage', and 'myth';

- discuss tourism as a part of globalisation; (and)

- discuss the role tourism can play in national social development.

Introduction

This chapter is written in a slightly different style from the previous ones. At times it necessarily follows a sort of stream of consciousness flow rather than a formal structure; this helps capture the flavour of the arguments and paradoxes and emphasises the fluidity of debate. It is important to understand from the outset that in many cases there are no right or wrong answers to questions concerning tourism's social dynamic. There may be an ethical element or a business element that will seem 'right' to their various proponents, but both standpoints are ridden with value judgements. Reading some of the earlier academic writings on the anthropology of tourism (such as Turner and Ash, 1975; George Young, 1973), there is sometimes an overwhelming sense of the authors 'judging' the tourist to be 'wrong' or even boorish: cultures continually change, even when tourism and tourists are not present.

Even so, tourism is certainly associated with change, but demonstration of an association does not necessarily mean that tourism has actually caused that change. Association and causality are not inevitably the same thing. Tourism is not automatically a main cause of change, but only one of a number of channels for the transmission of new ideas. It would be equally naïve however, to deny the role that tourism may have in precipitating or accelerating rapid change. This is where the need for more anthropological field work becomes evident, especially over long-term periods of time (so-called 'longitudinal studies').

All these ideas feed into the need for a snapshot of the broad themes to be found in the anthropology of tourism. These are shown in Figure 6.1.

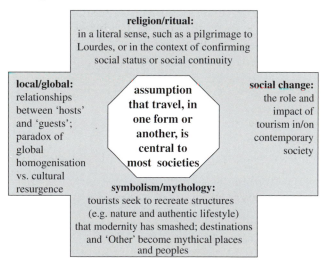

Figure 6.1: Some broad themes in the anthropology of tourism

For the present purpose, the issues in each of these themes may (if somewhat simplistically) be captured by a single question:

- Tourism as religion and ritual: *is tourism a modern form of religion or pilgrimage?*
- Tourism as social change: *does tourism damage culture?*
- Tourism as symbolism and mythology: *can tourism offer 'Paradise on earth'?*
- Tourism in local–global relationships: *will tourism bring development?*

Each of the themes will now be discussed in more detail.

TOURISM AS RELIGION AND RITUAL: IS TOURISM A MODERN FORM OF RELIGION OR PILGRIMAGE?

Pilgrimage may be defined as 'travel to sacred places undertaken in order to gain spiritual merit or healing or as an act of penance or thanksgiving' (Hoggart, 1992:236). A formal definition will also include the idea that pilgrimage must embody certain steps which are along the lines of the stages of a rite of passage: 'the start of the journey; the journey itself; the stay at the shrine or site where the sacred is encountered; (and) the return home' (ibid). It can be seen that there is also similarity here between certain types of leisure consumption which may be framed by:

- a stepping aside from normal rules of life and society;
- limited duration;
- unique social relations (for example role reversals, mixing of classes, rapid making of 'friends'); (and)
- feelings of intensity (perhaps pleasure or sensuousness).

If a slightly broader meaning is given to the term pilgrimage, if we include the idea of paying tribute at cultural stations rather than religious ones, then tourists may progress (in a social and intellectual sense) from their leisure consumption to a range of benefits similar to those gained by devout followers of a religion. Visits to the Eiffel Tower, MGM studios, the Tate Gallery and Disney World may well be of considerable cultural importance and follow at least one characteristic of Durkheim's notion of religion's essential role in social cohesion.

Nevertheless, not all travel has such profound meaning. If the travel and the on-site experience is largely activity orientated, it could be the case that there is no chance for van Gennep's sacred 'away-ness' of the those who travel away from home. The typical package holiday could be an excellent example of a trip with limited, or no overt symbolic meaning. The tourists have few opportunities to establish their own itinerary. They temporarily inhabit a world where the spatial and temporal rules are set by others, notably the travel company rep or tour guide. Some tourists may not exhibit any role reversals or indulge in feelings of intensity. Apart from a few obvious changes to dress modes and eating patterns, they behave (except for 'not being at work') as they do in their everyday life.[1] The experience, thus, may well remain secular: the tourist never finds himself as a stranger, on the contrary, he remains surrounded by other members of his 'tribe' of visiting tourists. In these circumstances it is only with the utmost academic endeavour that tourism can be seen as a form of ritual journey which is undertaken with (to reiterate Hoggart's definition) the intention of gaining spiritual merit or healing or as an act of penance or thanksgiving. This emphasises the twofold point made by many academics:

- that it really is difficult to get an agreed definition of *tourism*; (and)
- it is equally difficult to define the *tourist* to everybody's satisfaction.

The point we arrive at repeatedly is that it is wrong to assume that there is one single generic 'tourist-type'. While Cohen (1972) and others have been saying this for at least two decades, it has not always fed into the academic reflections that enter the public domain as journal articles or books.

So, is tourism and pilgrimage the same thing? Not so, say Boorstin (1964) and Barthes (1984). For them, tourism is an inauthentic pseudo-event (cf. Chapter 5) characteristic of capitalistic society in the mid to late twentieth century. However, MacCannell (1976); Nash (1981) say *yes*: they are both a search for authenticity in i) Self and ii) Other.

Cohen (1971), in an article called 'Arab Boys and Tourist Girls in a Mixed Arab/Jewish Community' says *yes* and *no*! The article draws on the idea that there are various modes of tourism in relation to what tourists might see as central to their being (the 'cosmos'). This might be recreational, diversionary or experimental. All three of which relate back to, or are framed by reference to the tourist's *home* as centre of life (cosmos). Under these circumstances, tourism can't be both strongly connected with home and a pilgrim, thus *no*, tourism is not pilgrimage! Only true pilgrims are committed to a cosmos external to their native society and culture. However, the role of the tourist can be combined with that of pilgrim, and an individual may feel a sense of belonging to more than one cosmos, thus 'yes' tourism is a form of pilgrimage!

Brown (1996) is rather clearer. He says that 'tourism and pilgrimage are opposed aspects of a single mode of interaction, between the travelling tourist/pilgrim and those whom he encounters on his travels (the so-called "guests" and "hosts")' (1996:44). On balance, it can be reasonably argued that there is no clear division between tourism and pilgrimage.[2] This is generally supported by: Turner and Turner (1978), *Image and Pilgrimage in Christian Culture*; Passariello (1983) 'Never on a Sunday? Mexican Tourists at the Beach'; Graburn (1983), 'To Pray, Pay and Play, The Cultural Structure of Japanese Domestic Tourism'.

Summary key points

- Tourism may be considered as a form of pilgrimage in the sense that it may mirror similar stages or characteristics (i.e. a ritual journey from the ordinary state to the spatially separated non-ordinary state for a set time);

- tourism offers release from the ordinary, routine life and that sometimes (but not always) this release includes freedom from social constraints;

- tourism can offer a chance for self-reflection and personal transition (with the bonus that knowledge and understanding of Other can endow societal respect); (and)

- more and better field research (including ethnographies of tourist groups) is needed to give a much stronger empirical base for the assertions about tourism as pilgrimage.

Questions

1   In what sense are modern forms of tourism 'sacred'? *Aren't we stretching and modifying the meaning in applying it beyond the point of usefulness?*

2   What are the rites of passage that occur in post-industrial/post-modern societies? *To what extent is tourism necessary in meeting this need?*

3   In a highly mobile world, both socially and geographically, *are we in danger of imposing a very narrow and limited concept (stages of rite of passage) on a wide and heterogeneous collection of different events?*

4   Are all touristic events in a capitalist society 'pseudo' and 'inauthentic'?

5   Is being a tourist a *behaviour*, a *status*, or a *state of mind*?

TOURISM AS SOCIAL CHANGE: DOES TOURISM DAMAGE CULTURE?

The encounter between 'host' and 'guest' is of profound importance in the study of tourism. At least two main themes occur. First, a range of cross-cultural interactions which becomes of heightened significance for our purposes when there is disparity between the visitor and visited. Second, there is a range of arguments surrounding the notion of 'hosts' and 'guests'. The key point here is that the words are used in an *ironic* sense, the special rules that apply to willing hosts receiving invited guests in their home are suspended. The transaction becomes a commercial one. A problem arises when expectations have been raised:

- on the side of the 'host' where government campaigns have stressed or overstated the direct economic benefits arising from tourism; (and)
- on the part of the 'guest' who may have been exposed to exaggerated advertising literature from tour operators promoting the 'friendliness of the natives'.

The second theme concerns the concept of 'strangerhood'. These touristic encounters, as Cohen (1972) and Nash (1977, 1981) have described, involve a relationship between strangers coming from different cultures or subcultures. The complex nature of these interactions will vary according to several factors, such as:

- the type of tourists (differentiating, as Cohen (1972) does, between *institutionalised* and *non-institutionalised* forms of tourism);[3]
- their length of stay, attitudes and expectations (which will affect their capacity to make relationships);
- the number of tourists (in the sense that fewer numbers means that tourists remain a novelty and with increasing numbers become just part of the scenery);
- the length of season (which will affect the local employment and give periods of 'rest' from tourists); and
- the role of the 'culture-brokers' or 'marginal-men'.

This last category is an interesting one. Culture-brokers or marginal-men (Smith, 1977) are defined as multilingual and innovative mediators that can control or manipulate local culture for tourists' purposes. Their role is often crucial in setting an entrepreneurial context for tourism development. As Nuñez noted:

> the acquisition of a second language for purposes of catering to tourists often results in economic mobility for people in service positions: interpreters, tour guides, bilingual waiters, clerks and police are often more highly compensated than the monolinguals of their community.
>
> (1989:267)

Culture-brokers can introduce change within their society. During periods of rapid and stressful change, these marginal men, being less conservative than the traditional leadership and perhaps more imaginative, may assume positions of leadership and may become successful innovators. In other words, culture-brokers develop certain levels of control over the amount and quality of the communication between hosts and guests. According to Mathieson and Wall 'they are in a position to manipulate local culture for tourist purposes without affecting the cultural identity of the host society in a detrimental manner' (1982:163).

Culture-brokers notwithstanding, the paradox that occurs here is the general assumption that a meeting between 'hosts' and 'guests' will take place.[4] Pressures arising are illustrated and further elaborated in Figure 6.2.

Opportunities for inter-cultural interactions may be limited if the distribution and planning of the tourist resorts is in the form of isolated or semi-isolated enclaves. The relatively small number of encounter possibilities is restricted to an insignificant part of the local population, mainly those directly or indirectly employed in the tourism industry and perhaps to those living in the surrounding area or close to special tourist attractions and places of interest.

Another important aspect of this limitation is the fact that tourists' interest in being in contact with locals may not be the main

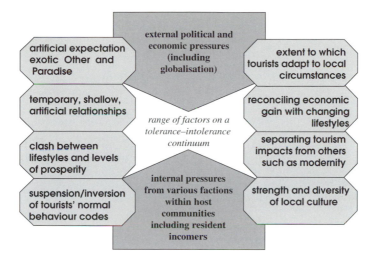

Figure 6.2: Pressures on host–guest relationships

objective of their vacation, on the contrary, they may be seeking leisure, pleasure, and escape from everyday cares and normal patterns of life.

Four other areas of host–guest relations need further explanation at this stage. They are: the demonstration effect; internal or external change; acculturation and cultural drift; (and) cultural symbiosis and assimilation. They are examined in turn.

The demonstration effect

This effect refers to the process by which traditional societies, especially those who are particularly susceptible to outside influence such as youths, will 'voluntarily' seek to adopt certain behaviours (and accumulate material goods) on the basis that possession of them will lead to the achievement of the leisured, hedonistic lifestyle demonstrated by the tourists. In this case, the tourists act as inappropriate role models for an unrealistic lifestyle. However, empirical evidence for this is somewhat weak and superficial. With the coming of what globalisation theorists call 'the compression of the world' (Robertson, 1992) it has become almost unrealistic to attempt differentiation of social impacts caused by tourism and the general process of modernisation. Where such impacts have occurred in the past, they have been derived from the following:

- misconceptions and stereotyping about the tourist. In the local's eyes, they may appear both wealthy and indolent; (and)
- the development of an inferiority complex in the local community which sets off a process of imitation. The apparent material superiority of the visitors is seen as desirable and as a model of lifestyle to be copied.

In these particular circumstances, tourists can arouse jealousy. Their values, behaviour and spending patterns are emulated and adopted by the host population. And, in the case of economic disparity between host and guest, expectations are raised unrealistically. In summary, two types of results can be observed in locals. First, changes in the value system, attitudes and language, and second, changes in dress, eating habits and demand for consumer goods (Mathieson and Wall, 1982).

Internal or external change

As we have established very clearly in Chapter 4, there are no cultures that remain static and unchanged over time. Cultural change is induced by two processes:

- *internal* by evolution through invention, driven by necessities or capitalism; (and)
- *external* through changes forced by outside economic, political, environmental and cultural influences.

Internal changes involve creativity, invention and innovation and are thought to occur at a more rapid pace in technologically complex and consumption-oriented societies[5] than in less-developed ones. External changes through the process of modernisation or

Westernisation, and external pressures have induced the greatest amount of cultural evolution, tourism being only the channel. No community or society is immune from outside contact, but the tourist is more ubiquitous than any other kind of representative of other cultures (Nuñez, 1989). The process of infusion demonstrates the following characteristics:

- selectivity (items of material culture are more strongly infused than the ideas and behaviour patterns);
- reciprocity (involving a two-way borrowing);
- transformatory (involving reinterpretation of cultural process and production in form or function); and
- dependent on a number of less clearly defined variables (such as the duration and intensity of contact, and the degree of cultural integration).

While tourism is clearly one of a range of catalysts for change, its ubiquitous nature means that it has the added capacity to profoundly affect the host community. However, the prevailing force remains the inexorable power of modernisation. Freda Rajotte wrote this with regard to the Pacific island nations:

> The most evident impact of tourism is in intensifying the change from a primarily non-monetary subsistence agricultural economy to wage-based, profit motivated tertiary activity. The profit, accumulation and investment motivation of large scale tourism development is in radical conflict with the kinship based, sharing ethic of many Pacific islanders.
>
> (1980:8)

Acculturation and cultural drift

Cultural drift is characterised by a temporary transformation in the hosts' behaviour, only for the duration of the encounter or interaction between hosts and guests (a *phenotypic* change). When the contacts are more permanent and continuous and a change happens to the norms, values and standards of hosts from one generation to another, we talk about a change in the *genotypic* behaviour. Therefore, the aspect that interests tourism researchers is whether the change is temporary or permanent. Acculturation is the process by which a borrowing of one or some elements of culture takes place as a result of a contact of any duration between two different societies. Acculturation is said to happen when contact between two societies results in 'each becom[ing] somewhat like the other through a process of borrowing' (Nuñez, 1989:266). with the exchange of ideas and products. This exchange process, however, will not be balanced, but an asymmetrical borrowing, because the stronger culture (with no connotations of superiority) will dominate and begin to change the weaker (not meaning inferior) culture into something of a mirror image.

Language is a key indicator of the asymmetry of those relationships and of the extent of local acculturation. As Nuñez suggests, 'the usually less literate host population produces numbers of bilingual individuals while the tourist population generally refrains from learning the hosts' language . . . this, in a tourist-oriented community is usually rewarded' (1989:266).

This returns us to the theme of the culture-brokers or marginal-men and helps explain some of their success.

Cultural symbiosis and assimilation

Both these terms are derivatives of the concept of acculturation. They mean the replacement of one set of cultural traits by another (Spicer, 1968:21). They have been applied to both the effects and to the processes of change. In general, we talk about acculturation when it concerns the Less Developed Countries (LDCs). However, the terms cultural symbiosis and assimilation are usually used when referring to cultures of more or less equality (we are not talking about superiority here, but cultures framed by economic similarities and comparable access to a wide range of information).

Thus the term acculturation is most often used between rich and poor countries (or regions). The term 'cultural collision' has been discussed by Gee, Makens and Choy (1989) in their analysis of the impacts of tourism. They created a model of stages and effects of the evolution of inter-cultural interactions on local residents. It is very similar to Doxey's Irridex (1975). It goes like this:

1 Initially, there is what they term toleration or accommodation, when the (initially) limited numbers of visitors and locals co-exists in some kind of harmony.
2 The second stage is segregation, when social distance and separation affect the interrelationship by means of either avoiding or remaining confined to special tourist areas, 'the golden ghettos' of luxury hotels and shops that offer the 'comforts of home'.
3 This segregation turns to opposition, when tourists are rejected by members of the host population (who see them as rich show-offs) or vice versa, the hosts are rejected by the tourists (who see them as money-grabbing nuisances).
4 The last stage is diffusion (as mentioned above) when the two cultures begin, through a process of symbiosis and borrowing, to converge.

This model may have a superficial attraction, but the empirical basis is not sufficiently strong to develop a theory and it is yet another deterministic approach that suggests a pre-ordained way in which host–guest relationships change over time.

The implication derived from such models is that no alternative exists and that they can be applied to each and every cultural interaction. In these models, relations are predetermined and will pass though the same process arriving at the same end. Considerable objection can be made to this determinism, particularly to the last stage of diffusion. Although diffusion plays an important role in cultural change, diffusion or infusion do not necessarily come about as a result of a saturation of the level of tolerance of the hosts. Cultural phenomena such as diffusion and infusion can have both positive and negative implications. An example of a positive infusion can be found in Deitch's (1977) study of the native Americans of the south-west USA, where clearly the borrowing process did not come as a result of these negative reactions.

Another objection can be made to the term 'cultural collision' as this also implies a negative connotation arising from the interaction. Wood (1993) makes precisely this criticism when he attacks the idea that there exists some sort of original pre-tourist cultural baseline against which to measure tourism's negative impacts as being meaningless: tourists enter a dynamic context with many influencing factors of which tourism is just one. This is Wood reflecting on the problem:

> Some years ago in an early survey of the literature on tourism in South-East Asia, I lamented . . . that the questions posed in the literature often conjured up a billiard game, in which a moving object (tourism) acted upon an inert one (the local culture). *My complaint was that this treated culture as unitary, passive and inert.*
>
> (1993:66, italics added)

The added italics really summarise most of the arguments against the simplistic assertions about damage done to culture by tourism. This is in no way to let tourism off the hook, so to speak. Tourism is very clearly a powerful force in every sense). But, tourism as an economic and cultural phenomenon presents a soft target to 'blame' for any cultural ills that befall a community. It has the added attraction of taking away any self-criticism or even responsibility! 'Look, it's all their fault, it's the tourists, not us!' Wood goes on to make another important observation:

> International tourism neither 'destroys' culture nor does it ever simply 'preserve' it. It is inevitably bound up in an on-going process of cultural invention in which 'Westernisation' is probably in most cases of lesser importance than other new directions of cultural change. It has its own peculiar dynamics which make it an interesting and challenging field of study, but

tourism's impact is always played out in an already dynamic and changing cultural context.

(1993:67–8)

Summary key points

- Relationships between hosts and guests and how they are formed and changed over time is of profound importance to the anthropological study of tourism;

- the nature of these relationships will vary according to a number of factors including the number and type of tourist and the relative economic status of hosts and guests;

- the ability to speak the language of the tourist (among other things) has produced what Valene Smith (1977) and others have termed 'culture-brokers' or 'marginal-men' who may influence host–guest relationships at a destination;

- stereotyping on both sides of the host–guest relationship can cause misunderstandings and superficial feelings of superiority or inferiority; (and)

- while socio-cultural phenomena such as the demonstration effect, acculturation, cultural drift, cultural symbiosis and assimilation have certain value, they are by no means 'proven' by empirical evidence such as hypothesis-testing or field research.

Questions

1   Do tourist brochures play any part in the relationships and expectations between 'hosts' and 'guests'?

2   Are independent travellers *better* than 'institutionalised' tourists?

3   Describe the role of taxi drivers as 'culture-brokers' or 'marginal-men'.

4   Do arguments about the 'demonstration effect' of tourism stand up in the light of global information and communication technologies?

5   Discuss the internal and external changes to a community with which you are familiar. (These changes may be general ones, not necessarily touristically induced.)

6   What are the arguments against the notion of 'cultural collision' in the context of tourism?

TOURISM AS SYMBOLISM AND MYTHOLOGY: CAN TOURISM OFFER 'PARADISE ON EARTH'?

In David Lodge's ironic novel, *Paradise News*, one of the principal characters is Roger Sheldrake, a tourism anthropologist. His central contention is that:

> sightseeing is a substitute for religious ritual. The sightseeing tour as secular pilgrimage. Accumulation of grace by visiting shrines of high culture. Souvenirs as relics. Guidebooks as devotional aids . . . I'm doing to tourism what Marx did to capitalism, what Freud did to family life. Deconstructing it.
>
> (1991:74–5)

Later on we find the intrepid anthropologist engaged in his field research at the 'Wyatt Regency' hotel in Hawaii:

> He took a notebook out of his shirt pocket and ran through the list: 'Paradise Florist, Paradise Gold, Paradise Custom Packing, Paradise Liquor, Paradise Roofing, Paradise Used Furniture, Paradise Termite and Rat Control . . .' He spotted these names on buildings or sides of vans or in newspaper advertisements. He is working on the theory that the mere repetition of the paradise motif brainwashes the tourist into thinking they have actually got there.
>
> (1991:163)

Contained within this fictional account is the germ of an idea. It takes Graham Dann (1996) to do it justice. His analysis of tourist brochures, along with two earlier works by Uzzell (1984) and Selwyn (1993), *deconstructs* (using the fictional Roger Sheldrake's approach) brochures in their own right: analysing the language and imagery of them and the 'consumerist myth' (Selwyn, 1996:16) contained therein. Dann studied over 5,000 images featured on almost 1,500 pages in 11 British holiday brochures. He was able to establish four types of 'paradise' to be placed before the consuming public of potential tourists:

- Paradise contrived: *no people; natives as scenery; natives as cultural markers*;
- Paradise confined: *tourists only – tourist ghetto*;
- Paradise controlled: *limited contact with locals: natives as servants, natives as entertainers, natives as vendors*; [and]
- Paradise confused: *further contact with locals, attempt to enter locals-only zones: natives as seducers, natives as intermediaries, natives as familiar, natives as tourists, tourists as natives.*

(1996:68)

Dann's main point is that 'less than 10 per cent of the [photo images] were tourists and locals shown together, an indication that, for the media-makers at least, the idea of tourism as a meeting of peoples was somehow not to be encouraged' (1996:64). Here we see then the commercial brochure either as a sop to existing fears and prejudices of potential tourists, or more complexly, as a deliberate construct setting out the rules of play before you go. In either case, it is a far cry from the explicit message of 'enjoying local culture' to be found in the same brochures: thus the brochure becomes, as Selwyn puts it, 'an instrument not for greater democracy but for greater social and political control' (1996:16).

To add my own contribution to this discussion of brochure-paradise, in the middle of Virgin Territory's 1994 holiday brochure *The California Collection*[6] under the headline 'In Search of

Paradise', the Pacific island nation of Western Samoa is described as being 'impossible to do justice to within the context of a short introduction' but that:

> if your dream of the far Pacific Isles is one of smiling people, of warm sunlit islands where time has stood still, then Western Samoa will not disappoint you. Robert Louis Stevenson lived, loved and died here . . . You will find no 'commercialism' in Samoa. Amazingly our 'Western Lifestyle' does not prevail. Crime is virtually non-existent and strangers are welcomed with legendary hospitality. In 'town' you will be greeted with a smile but in villages you will be welcomed with pride like a returning family member. For each village is one extended 'family' and your presence will be an honour for that family.
>
> (Virgin Holidays, 1994)

These generalities are wrong. They are at odds with the complex social rules whereby a Samoan family will 'adopt' a visiting stranger. But this is to protect that stranger on the basis that not to have a family is to be a nobody, a situation described by the travel writer Paul Theroux in the following terms:

> having a local family gave you status and protection. Samoans quite freely co-opted strangers and made them part of the family . . . if you were alone on the islands and did not know anyone you would be victimized . . . The family looked after itself but was indifferent to the plight of other families.
>
> (1992:335–6)

Virgin Holidays' description of Apia (Western Samoa's capital) as 'the most charming and unspoiled in all Polynesia' contrasts richly with Theroux's perception of the town which he describes as:

> the squalid harbor town . . . [which] seemed to me mournfully rundown, with broken roads and faded and peeling paint on its

ill-assorted wooden buildings, and Samoans rather gloatingly rude and light-fingered, quoting the Bible as they picked your pocket.

(1992:324)

While Theroux's well known cynicism means that descriptions by him should be taken with caution, his work none the less provides a powerful contrast with the vacuous descriptions and market-oriented blandness found in the tourist brochures.

In a broader sense, the question of constructing paradise is just one part of a range of pressures on culture through tourism. The myths of tourism extend far beyond the creation of paradise through brochures. Tom Selwyn's book *The Tourist Image: Myths and Myth Making in Tourism* (1996) shows us that it extends into the imagery of postcards that help construct Other for touristic consumption, the power of museums to construct the past by way of a 'common heritage' that never existed, the way in which just 'walking through nature' can generate feelings of nationalism and nostalgia for imagined places, and (if we switch to Jacqueline Waldren's book on *Paradise and Reality in Mallorca*) we discover that:

Paradise and reality flow into one another and the ills of the outside world are temporarily closed out by recalling exciting moments and beautiful memories through 'myths' of Deià's [her adoptive village] past [. . .]

The outside world becomes more and more distant as the inside becomes larger and more complete.

(1996:200–201)

Myths of one sort or another are found in all societies. As 'rational science' tears away our traditional myths of dragons, princesses and creation for being fallacious or unfounded, so we invent new ones about 'Diana, Goddess of Landmines' or 'the time before Rock and Roll when we could leave our doors unlocked' or 'the Swinging Sixties' or that most potent myth, 'Free Speech'.

A previous generation of anthropologists spent a great deal of time trying to differentiate mythology from history.[7] It is generally accepted now that they serve the same function in society, it is not so much the content of the myth that is important but the structure, for it is this, as Lévi-Strauss has it, that reveals the universal mental processes.

Summary key points

- The myth of paradise remains an important factor in the marketing of destinations;

- the danger arising from this is that social cohesion breaks down when myths are shattered, so people invent new myths: this might include the myth of the 'stupid local'; (and)

- hosts and guests create myths about each other and these myths can frame the temporary relationships that arise.

The final word of summary must go to Tom Selwyn who says this in his introduction to *The Tourist Image*:

> Tourist perceptions, motivations, and understandings about destinations are shaped by a preoccupation with harmonious social relations, ideas about community, [and] notions of the whole. These are the preoccupations which are mythologized in the tourist's view of Nepal, the English West Country or wherever – and which are, in this sense, 'overcommunicated'. What is concealed, however, (by the forgetfulness that mythical language permits) are the actual fractures and displacements which [pragmatic and empirical case studies] clearly reveal.
>
> (1996:3)

Questions

1 'Tourism is about play and not cultural politics.' Discuss this from the perspective of i) the tour operator, ii) the consumer (tourist) and iii) the local community.

2 Take a current tour operator's brochure and i) decide on the particular tourist type and market segment it is aimed at (presenting evidence to back your claim) and ii) analyse the text and images to tease out the hidden meanings.

3 Does tourism contribute towards inter-cultural understanding?

4 What are the advantages and disadvantages of using models such that proposed by Doxey (1976, cited in Mathieson and Wall, 1982) to analyse tourist–resident tolerance and relationships?

Tourism in local–global relationships: will tourism bring development?

In a sense, this section is about the political economy of tourism. Within a social science analysis of tourism, and picking up on Lévi-Strauss' notion of 'symbolic dualisms' (cf. Chapter 1) a recurring pattern of bi-polarities is evident. This is shown in Figure 6.3. This allows for an overview of the tensions and contradictions that are seen in discussions about tourism and which unfold during the narrative of this book.

Some of these symbolic and economic oppositions will have been seen before, arguments about tourism as a system and the interconnections between tourism and culture were dealt with in Chapters 2 and 4 respectively, while issues of globalisation and development will be discussed in Chapters 7 and 8.

However, I would argue that it is precisely because tourism as an 'industry'[8] does impact on people's lives, that anthropology can play an essential role in drawing together some of the binary separations illustrated above.

global focus	local focus
economic enlargement tourism-as-industry tourism-as-consumerism globalisation core modernisation	sustainable human development tourism-as-system tourism-in-culture/culture-in-tourism localisation periphery underdevelopment
aiming to maximise market spread through 'familiarity' of product; homogenised, undifferentiated product dependent on core; focus on tourism goals as defined by outside planners and tourism industry	*aiming for independent, differentiated destination with decreased dependency on core; focus on development goals as defined by community; role of local social institutions*
individualistic	**holistic**

Figure 6.3: Bi-polarities in tourism's global–local relations

It can be seen that some forms of tourism, such as those sold as packages or which involve high levels of resources, are an integral part of the global political economy. It is this characteristic that contributes to the *structural* nature of the 'tourism-as-development' problem, and makes it interesting for anthropologists as observers of how global–local relationships work. It is the tourism system(s) in place at a given destination, including the unobservable social and economic *structures*, that marginalise(s) potential participants in the industry (such as those in rural areas or who are not part of an élite) and is thus capable of delivering *growth* (in an economic sense) without *development* (in a societal sense). This is the sort of tourism seen on the left hand side of Figure 6.3, the global focus.

This need not be the only system for tourism. For some locations and tourist destinations, an alternative approach is needed along the following lines:

- the development of a tourism system enriched by anthropological field work and observations of relationships;
- the recognition of the necessity to develop social structures and institutions (including NGOs); (and)

• the contextualisation of these two aspects firmly in the development process.

Such alternative approaches would help place 'ownership' of the development process along more equitable lines. Global–local relationships that are led only by a thirst for tourism development (as opposed to tourism as cultural phenomenon, or tourism as having a potential for social growth) skews progress (but not economic enlargement) because they place prime importance on the interests of the product-suppliers and travel-intermediaries, rather than on a broad spectrum of social institutions that will have a wide range of interests in development.

The paradox arising from global–local relations framed only by a profit motive is that they fail to recognise that in the long-term, this one-sided approach, which sees the development of a range of social institutions (including NGOs and community level organ-isations) as being of secondary importance, will cause civil society to implode, witness Algeria and parts of Upper Egypt (Aziz, 1995).

It can be seen from the above that the overarching problem is that of confusing the various debates that surround tourism's issues. While they are clearly inter-linked (for example discussing global environmental impacts without referring to local impacts on culture is untenable) it is confusing to intermingle the debate about the worth of tourism as a *development tool* with the discussion of tourism as *'needs satisfier'* (to borrow from the language of marketing). These are separate arguments. Yet it is this type of confused message that drives many of the élitist detractors of tourism who make such statements as: 'truly aware people would not go on holiday at all. If we took more time and resources to make *here* look more like *there*, to be *here* would become more attractive!' (Nicholson-Lord, 1990:5), and Sewell who describes tourists as being 'nincompoops . . . who would, in general, be far happier boozing and burning in the sun, and for whom ancient cathedrals . . . are mysteries beyond comprehension' (1996:12).

Thus we see a range of arguments against the notion of 'tourism from above'. This is not an anti-tourist message, but as tourists become more restless and seem less contented with what is offered to them (Poon, 1993) and alongside this we have the parallel development of increased expectations on the part of host populations, then the existing model of tourism (i.e. supply-led and intermediary-dominated) cannot continue without radical surgery: it is simply not sustainable.

The World Tourism Organization predicts the doubling of tourism by the year 2020, but offers no explanation as to how the impacts on global–local relationships are to be tackled. The need for resolution is as much for the well-being of the tourism 'industry' as it is for the welfare of the host communities who will bear the brunt of the rise in tourists' arrivals. The kind of solution

Protect	what is natural and beautiful for the benefit of 'natives' and tourists
Reduce	density – do not overcrowd an area with too many hotels, tourist shops, or visitors
Enhance	the feeling of seclusion and privacy to preserve the feeling of the area and the psychology that travellers want, i.e. a retreat and escape from the cares of the world
Seek	quality throughout since it entices visitors to stay longer, brings them back time after time, and ensures a longer life for the area
Emphasise	diversity – the more activities that are available, the greater the likelihood that something will be of interest to each visitor each day
Restore	the natural and the historical, retain the sense of heritage, continuity and community
Value	local culture and traditions to protect local populations, their heritage and culture
Institute	height limits (buildings usually no taller than three storeys) to protect vistas and other scenic views
Negotiate	for open areas to provide the 'breathing room' that enhances every project
Gain	community acceptance so that the local populations are the ones who benefit most from tourism (this is their home), in partnership with commercial enterprises

Figure 6.4: The PRESERVING approach for a better tourism future
Source: Plog, 1994: 52–3

being offered by the World Travel and Tourism Council (WTTC) of a deregulated, self-governing enterprise is not a useful one. Neither is the current state of relationships between tourism academics and the 'industry' they analyse. Too often, academics are accused of being leftist élitists who are intent on rubbishing the industry, while the academics point the finger of blame at tourism executives who seem intent on subsuming all 'virgin' parts of the planet to their corporate bottom lines. Plog (1994:52–3) sets out an idealistic approach to forming new global–local partnerships which at least recognises that a problem exists. This is shown as Figure 6.4.

Summary key points

The first four points are linked, drawing together arguments about the contrast between rich and poor:

- the development of 'islands of affluence' in the midst of poverty;

- the use of scarce national resources for the enjoyment of wealthy foreigners;

- the 'demonstration effect' upon the local population of observing mass-consumption and indolence; (and)

- a general point about commoditisation or commercialisation of culture and lifestyles.

The last two concerns are to do with political economy and distribution of tourism's economic and social benefits:

- financial benefits are likely to accrue to foreign companies and local élites; (and)

- the realities of international tourism systems mean that control is likely to be external to the destination and defined by trans-national tourism corporations.

Questions

1 In what ways can tourism, which is surely about fun, be described as any part of an international political economy?

2 Explain the term 'undifferentiated, homogenised tourism product'. What is the significance of this term to local–global relationships?

3 What is the *structural* nature of the 'tourism-as-development' problem?

4 What are the long-term implications of a tourism policy based solely on economic benefits for tourism enterprises?

5 Explain the paradox of tourism as a *development tool* and tourism as a *'needs-satisfier'*.

6 What role can NGOs play in local–global relationships?

Key readings

One book stands out as presenting some fascinating insights into issues of the anthropology of tourism, that is of course, Tom Selwyn's edited volume *The Tourist Image: Myths and Myth Making in Tourism* (1996, Wiley). It applies an anthropological perspective to the study of tourist mythologies and looks at the relation between tourism, society and culture. Priscilla Boniface and Peter Fowler's *Heritage and Tourism in 'the Global Village'* (1993, Routledge) takes a look at the way tourism feeds off heritage, pleading for a better understanding of the global condition by those who have some sort of management responsibility for leisure, heritage and conservation. In a more general sense, John Corner and Sylvia Harvey's edited book *Enterprise and Heritage: Crosscurrents of National Culture* (1991, Routledge) is useful. The chapters draw on aspects of culture such as film, television, urban planning, architecture, advertising and tourism to develop a critical appraisal of enterprise and heritage in British social and cultural life. Finally Graham Dann's excellent analysis of how language constructs aspects of tourism *Language of Tourism: A Sociolinguistic Perspective* (1996, CAB International) is essential reading.

7 Globalisation, tourism and hospitality

Plate 7: Whatever decisions are made at a global level concerning investment and marketing, at some stage the local population will have to be mobilised to engage in the business of tourism. These are young Samoans receiving training in tour guiding. Emphasis was placed on giving tourists accurate information about Samoa and its traditions. Hospitality to strangers is governed by patterns of gift exchange that are not always understood by all the players. Employment in tourism is seen by the Samoan government as one way of combating outward migration to New Zealand and Los Angeles. Anthropology is about people: increasingly the relationships between people are framed by global influences including advertising and cybermedia.

Overview, aims and learning outcomes

The aim of this chapter is to:

- introduce some of the arguments of the globalisation debate;

- place tourism (and hospitality) within a specific global context; (and)

- develop a critical perspective on globalisation.

After reading the chapter you should be able to:

- understand that the notion of an emerging 'global village' is a contentious one;

- evaluate, at a preliminary level, the extent to which governments are losing power over trading and economic relations in the face of globalisation; (and)

- describe, at a basic level, how the general process of globalisation is affecting international relations.

Introduction

In a general sense, it was probably the 'ozone' debates of the mid-1980s that first brought home to the general public that actions taken in one part of the world can have dramatic effects on other parts. These can be illustrated by such examples as the rise in mercantile 'flags of convenience' and the links to accidental sinking of oil tankers,[1] or Chernobyl, or the acid rain from British industry causing the death of forests in Sweden. Global warming, with its attendant conferences and debates illustrated that there was more to global theory than global products 'borne of a high-tech, fast-moving society, frequently allied with the motive to maximise profit' (Boniface and Fowler, 1993:3). It is a phenomenon that encompasses both culture and communications.

Globalisation theory

The term 'globalisation' is rooted in the study of international relations and 'modernisation'.[2] It has been summed up by Anthony Giddens as 'the intensification of world wide social relations which link distant localities in such a way that local happenings are shaped by events occurring many miles away and vice versa' (1990:64). Ankie Hoogvelt puts it in more dramatic language 'globalisation today is essentially a *social* phenomenon that drives cross-border economic integration to new levels of intensity . . . Globalisation is a *process* not an end-state of affairs' (1997:131).

It differs from the simplistic proposition that we all live in a world which, through global products and consumerism, is increasingly the same. There is a line of thought that says that far from nations moving, at different speeds, towards the same end, 'industrial man' as Robertson (1992:11) terms it, which implies a sort of eventual equality, there has been a far more complex economic and cultural shift. In some respects, this is typified by the dominance of Western style management and global corporations based in metropolitan centres of capital and knowledge power. Robertson indicates that globalisation might be characterised as 'compression of the world' (1992:166). This can be linked to Marshall McLuhan's concept of 'the global village' (1960). Dunning also identifies this trend, 'as the cross-border interchange of people, goods, assets, ideas and cultures becomes the norm, rather than the exception, so our planet is beginning to take on the characteristics of a global village' (1993:315).

This is one view. It is no longer a useful one. It may be argued that for the most part, the tendency towards a 'global village' (even when used as metaphor for instant communications) is almost solely the territory of élites. Those that have access to the electronic superhighway which is not economically or politically available to large parts of the globe. Thus we can imagine that rather than a global village, we might have a series of 'manor houses' (to use a metaphor from feudal times) where like-minded élites commu- nicate with each other across the globe surrounded by seas of poverty inhabited by those who don't communicate outside their own reference groups and who, with their limited access to quality information, don't know, to coin a phrase 'what is going on'. Even with the electronic revolution, there are still parts of the globe which remain, as Robertson tells us 'uninformed and lacking in "adequate" and "accurate" knowledge of the world at large and of societies other than their own (indeed of their own societies)' (1992:184).

Perhaps Dunning's and McLuhan's concepts ought now to be termed global fiefdom, with dominant economic power being divided into a 'triadic division of the contemporary world'[3] (Robertson, 1992:184). This raises the issue of motivational clash: the economic versus the cultural, i.e. where world trends are being dictated by MNCs with their zeal for free markets and profit maximisation, while populations want to retain their identity with efforts to reject the homogenisation of their culture.

Here, however, a paradox arises for globalisation theory: in the face of such power-blocs, there is also increasing nationalism, fragmentation and polarisation, for example within an intra-power bloc, demonstrated by the political situation in the Balkans, and an inter-power bloc such as the continuing war of nerves over trading between the US and Japan. The technological revolution in global communications, still epitomised by the CNN coverage of the Gulf War – especially from Baghdad, has not done anything to increase understanding between nations. CNN cannot be held accountable for this. Their mission, as a commercial broadcasting corporation, is to increase the number of viewers thus increasing their capability to generate income through advertising revenue. Political action (or more accurately *re*-action) in the light of famine, war or other local events is increasingly being defined by what hits national TV news.

The process of globalisation

The general process of globalisation has been driven by the MNCs which are powerful enough to develop and implement global strategic business plans framed by cost-reduction measures. The rise in a globalised economic order has accelerated from the period of the mid 1980s when there was a switch away from centrally administered economic policies towards market oriented strategies: a phenomenon commonly known as Thatcherism or Reaganism. The rise of MNCs has had an effect on a nation's ability to control their private sector businesses. Ozay Mehmet puts it this way, 'MNCs are no longer "at bay". On the contrary, in the new world economy, the MNCs are "in" and the nation state is "out" as sovereignty is increasingly eroded by the forces of globalisation' (1995:130).

This argument of Mehmet's is a crucial one. Central to this chapter is the notion that economic power is being taken away from national governments leading, in effect, to lessening of control over domestic policies concerning social, employment and environmental issues.

The GATT round in 1996, (in which GATT was effectively replaced by the World Trade Organization)[4] actually weakened the power of governments to set standards and control their own economies. The 'new' WTO is a major institution equivalent in power to the World Bank. It will have veto over decisions relating to the regulation of commerce and the setting of labour, health and environmental standards up until the present within the competence of democratically elected bodies. The implication is of a convergence towards a sort of global common standard, determined by the 'new' WTO, of national laws, regulations and administrative procedures which encompass trade and environmental, health safety and labour standards. A country will be able to challenge the laws of another if it believes that 'new' WTO objectives of so-called 'free trade' are adversely affected. Laws can also be challenged if they aim to protect or support local industry.

It is almost as though this 'new' WTO will determine global economic policy. Such an entity is outside the United Nations and accountable only to its board of directors. No doubt, the board will be fully under the influence of its Western proponents. We can see this happening in the case of American pressure on the EU to stop its historic special relationship with banana growing nations of the Caribbean. The argument is said to be about free trade and access to markets for the South American growers. Not so. It is an example of US big government supporting US big business (in this case, Del Monte, Dole and United Brands). This is having a devastating effect on the economy of the Windward Islands. Thus we discover a political economy of bananas![5] Similarly, in July 1998, the British supermarket chain Tesco was prevented from selling brand names such as Nike, Levi's and Calvin Klein at rock bottom prices because of pressure by the manufacturers on the EU. They claimed that it introduced unfair competition by buying 'grey' imports from outside the EU.

A triumvirate of the 'new' WTO, the World Bank and the International Monetary Fund (IMF) seems destined to act as a sort of legitimising agent for the activities of the MNCs. Manufacturing, assembly and processing are strategically located and relocated around the global wherever current incentives are being offered by governments anxious to attract inward investment or where wages are low and employment laws 'liberal'. Technological advances and international pressure to eliminate 'protectionism' have strengthened the conditions for the rise of the global corporation: advances in communications, transportation, materials handling and data processing have made even industrial production almost portable. Thus, insofar as the Third World is concerned, it is not so much a *willingness* on their part to become part of this new global order, but an *inevitability*.

Tourism, hospitality and the Third World

For the tourism industry in general and the international travel trade in particular, another response to this synopsis is epitomised by the production of fake cultures and inauthentic images over which local people and communities have no say.

While it may be thought that as consumption and production of tourism products are locationally linked, mass tourism operators are quite capable of switching 'production' i.e. holiday products, to more profitable destinations should the need arise. In the case of one tourism sector, hotels, this globalisation process has been driven by a particular need, Go and Pine, suggest:

> to switch gears across geographies . . . [with] . . . the main purpose of a globalization strategy [being] to create and sustain value for all constituents thereby improving competitiveness . . . [framed by] . . . an evolving logic to respond to intensifying world-wide competition which drives customers' demands for value and leaves hoteliers little room for error.
>
> (1995:xvii)

Linking the above are the twin themes of Eurocentrism, as in the case of Western consultants dominating the development agenda (Burns and Cleverdon, 1999) and the actions of MNCs in shifting production via the process of 'social dumping'. For tourism and hospitality enterprises, this means the familiar cycle of searching out 'unspoilt' destinations, encouraging development, moving on when saturation (or novelty) point has been exceeded. The destination is left with an economy reliant on tourism, a workforce trained within a service culture and infrastructure oriented to supporting foreign 'play' – the options remaining open to such destinations are limited and their bargaining power lowered.

Globalisation for tourism enterprises

There is a particular problem for tourism firms following a global strategy. There exists something of a conflict between developing niche markets and globalisation: customisation versus standard-isation. The solution to the problem seems to lie with 'protecting' the customer from core, administrative elements that are best served through global policies (for a hotel chain this might be the standardised 'back of house' policies related to centralised inventory procedures, staff orientation policy and safety standards). Local management is then free to respond to the type of guest it receives.

Thus, while the company operates on a global basis, insofar as the guest is concerned, they receive 'customised customer care' at a specific location. The product then becomes a mix of core-operations that 'drive' the property and flexible response 'add-ons'.

The motivation for globalisation is compelling, especially for tourism enterprises whose product relies, in many cases, on an ever expanding geographic diversity. Technological advances in both transport and information systems have enabled these global service traders to think beyond traditional political–national boundaries. International competition has driven such corporations into thinking globally. By integrating corporate activities on a world-wide basis through referral systems, computerised reservation systems and vertical integration with other sectors, competitive advantage is gained by controlling the flow of tourists through the international distribution channels (i.e. with the various travel intermediaries).

As mentioned above, one of the continuing paradoxes for international corporations within the globalisation debate is that of standardisation. That is to say, standardising the 'product' which for our purposes may be taken to mean:

- the physical product such as architecture, decor and promotional literature;
- the intellectual product, meaning standard operating procedures and personnel practices; (and)
- the emotional product meaning 'hospitality' and service attitudes from front line staff in dealing with guests.

There is the simplistic argument that global markets require global, non-differentiated products, and marketing. The problem for tourism enterprises (as opposed to say, a manufacturing venture) is that part of the attraction is the unique character of the destination attractions. This tourism product mix may well include culture and people. This invokes an intellectual argument about whether culture and people should be enmeshed in corporate marketing strategies which is rehearsed elsewhere in this book. The process of standardisation, an almost inevitable consequence of the vast chains of connectedness that comprise the major tourism groups, may dilute what is most precious about a destination: its individualism – especially if it is being promoted to potential tourists in the industrialised and post-industrial world whose motivation for travel might well be a reaction to a lack of individuality. There are however, certain aspects of standardisation that could be implemented as efficiency measures such as 'single sourcing'. For example, Forte Hotels 'single source' all their international information technology provision from IBM (Jones, 1993:143). These issues will be dealt with at corporate level by the procurement division. Another argument for standardisation is that it assures consistency. This was (and remains to some extent) one of the founding tenets of the Holiday Inn empire and an idea which became almost an obsession in Holiday Inn's corporate thinking of the 1970s. Marketing and product were standard and global in spite of location or local conditions. In some instances this global strategy is clearly linked to financial success. The names McDonald's, Kentucky Fried Chicken and Coca-Cola conjure up a positive expectation of consistency in the products and confidence in the back of house operations (mostly to do with sanitation and safety issues).

While it is claimed that the drive for standardisation is becoming less of a priority in corporate strategic thinking (Olsen and Merna, 1993:95), the issue creates a paradox within the emerging work on creating customer-oriented organisations: a sort of cost-effectiveness continuum with 'reducing costs by standardisation' as one pole and increasing customer satisfaction through 'increased customisation' as the other. Bowen and Basch (1992:208) describe the greatest test for service-oriented organisations as being how non-routine enquiries or events are dealt with. The general move away from standardisation can be characterised by the following:

- greater understanding on the part of 'producers' that many of tourism's products are not 'naturally' homogeneous;
- customer reaction against 'industrialisation' of services, what

Ritzer (1993) generally refers to as McDonaldization of society; (and)

* realisation that the tourists themselves are not a particularly homogeneous group.

These themes have occurred in earlier chapters of this book, and do not need further reiteration, but it is useful to note the following impacts claimed for companies:

* greater decentralised decision making, *local managers taking responsibility and accountability*;
* moves away from rigid, hierarchical, mechanistic structures, *which may dissipate creative thinking*; (linked to)
* increased moves to organic management styles, *flexible interaction and lateral communication and knowledge sharing at all levels*.

However, for some organisations, particularly those suited to an industrialised process such as fast food restaurants, there is no mistaking the underlying trend not of post-modernism, post-industrialism or post-Fordism, but of modernism, i.e. corporate action firmly rooted in Taylorist 'modern scientific management'. The emphasis is on rationalism 'conveyor belt' work processes (not only in an allegoric sense, but also literally in that some ovens and grills literally work on the conveyor belt system). Thus we find a curious scheme of things where modernism prospers within post-modern society. In the final analysis, it is Ritzer (1993:157) who reminds us that so called post-industrialism and the con-comitant post-modern age is still defined by a capitalist mode of production which has been 'globally dominant for the past two centuries'.

Conclusion

The links between tourism, globalisation, and international politics is cultural on at least two levels. On the one hand there are the obvious cultural changes and connections that are a well acknowledged result of international travel. On the other hand there is also a growing awareness that perhaps the culture of governments is becoming 'the same' through the process of globalisation which, to be realistic, is out of their control, and as 'we are . . . in a period of globe-wide *cultural* politics . . . The "official" line between domestic and foreign affairs is rapidly crumbling' (Robertson, 1992:5).

This 'crumbling' division between the domestic and foreign at a national level has been accelerated through the catalyst of growth in global institutions. Movements resulting from global capitalism and global media systems have, as Robertson points out, caused the 'boundaries between societies [to] become more porous because they are much more subject to "interference and constraint" from outside' (Robertson, 1992:5).

It can be concluded that the part of international tourism which takes place in Third World destinations, which, for our purposes focuses on the actual holiday experience, does not come about through any sort of natural economic process within that country. Most Third World countries have little or no domestic tourism upon which to build their international tourism. Demand for any significant form of tourism (other than individual travellers 'passing through') is not even likely to come from potential tourists. Rather, the process by which such countries become a part of international tourism is likely to be through the specific actions of government, international tour and transport operators and foreign investors.

Key ideas

- Globalisation is a complex idea that encompasses both political economy and cultural ideologies;

- there are distinct elements of globalisation that are élitist by excluding the powerless and those without access to technology;

- GATT and the World Trade Organization (the 'new' WTO) does not necessarily work in the best interests of the Third World, and yet such countries are being inexorably drawn into a global system that is not of their making;

- hospitality and tourism firms face a paradox as they expand their businesses internationally in that on the one hand, they seek rational standardisation but the customer base is increasing their demand for differentiated products;

- there are different ways to examine globalisation. The two main approaches are the *sociology* of globalisation and the *economics* of globalisation; (and)

- globalisation means national borders become increasingly irrelevant, but at the same time a paradox arises demonstrated by the simultaneous rise in nationalism and neo-fascism.

Questions

1   Are people part of globalisation?

2   'There is no problem with either the "new" WTO or globalisation, there are just development problems.' Discuss this statement with reference to the Third World.

3   In what ways are bananas and beaches linked to the international political economy?

4   What are the particular dilemmas facing hospitality and tourism firms as they seek to 'go global'?

5   Is the world becoming a 'global village'?

Key readings

There are a number of books on the subject of globalisation. Some of them are dense texts that do not come across as user-friendly. However, as a basic introduction, Waters' *Globalization* (1995,

Routledge) is a useful text, with definitions and explanations of the terms used in each of the main arguments. Ankie Hoogvelt's book on *Globalisation and the Postcolonial World* (1997, Macmillan) provides a lively and at all times interesting explanation of what is happening and the historical context behind globalisation. Finally, Cynthia Enloe's *Bananas, Beaches and Bases* (1989, Pandora) is inspirational in its approach to analysing the problem of globalisation from a feminist perspective.

8 Charting development thinking

Plate 8: The 'promise' of tourism is economic and social development. The reality is often something else. Here, Sri Lankan street hawkers try to make a living selling postcards and trinkets in Kandi. Development is often targeted at inward investment so that large multinational corporations might get tax breaks and other forms of assistance whereas less attention is given to small scale entrepreneurs. The big question remains: 'Development for whom?'

Overview, aims and learning outcomes

The aim of this chapter is to:

- identify development studies as an important perspective on the anthropology of tourism;

- trace the history of arguments within 'development studies' as a field of knowledge; (and)

- establish a base upon which further, more detailed study is possible.

After reading the chapter you should be able to:

- develop a critical view on development problems and use this to revisit previous chapters on tourism and anthropology to begin establishing the importance of these in development;

- be aware that views and definitions of development will vary;

- understand that these viewpoints are framed by political economy; (and)

- describe, at a basic level, how the general processes of development is currently thought to take place.

Introduction

A central theme of this book is the relationship between tourism and development. In this chapter various perspectives on development are examined. They are all, in effect, about two issues that concern anthropologists. That is *growth* and *redistribution*. This sets a development context for the discussion of tourism, globalisation and the recurring theme of development that is evident throughout this book. As in the previous chapter, the style here varies from the earlier part of the book. This work on development is presented in a more formal manner, reflecting its rather heavy political grounding.

A special annex has been attached to this chapter which details the growth of 'development studies' as a field of knowledge. This is useful for identifying where the main arguments in the field have come from. However, the chapter can be read without this annex.

Describing development

> Development ought to be what human communities do to themselves. In practice, however, it is what is done to them by states and their bankers and 'expert' agents, in the name of modernity, national integration, economic growth or a thousand other slogans.
>
> (Adams, 1990:199)

The word 'development' is used to describe the dynamics resulting from processes of national economic and social transformation, 'effectively a synonym for more or less planned social and economic change' as Hobart (1993:1) observes. Hobart also argues the importance of the use of metaphors in the development discourse ('nature' metaphors such as roots and seeds, and spatial metaphors such as up/down, centre/periphery are examples he gives). In particular, Hobart cites three ways in which development is described in the Indonesian language:

> *Perkembangan* from . . . 'flower', . . . growth which requires little external intervention. *Kemajuan*, 'progress', tends to be linked to Western liberal economic and political ideas, with connotations of rationality. The third, *pembangunan*, from . . . 'get up, grow up, build' is the term favoured by government officials and developers.
>
> (1993:7)

Pembangunan being preferred by governments because it places importance on their role in the process: implying that development will not happen by itself, but needs carefully thought out government intervention. This political dimension to development supports Adams' (1990:4) assertion that development as a concept is a 'semantic, political and indeed moral minefield'. Todaro (1982) argues that development needs to be framed by the overall social systems and cultural dynamics of a country. This means taking a holistic[1] view of development and not one framed solely by economic enlargement measured, for instance, by the achievement of growth targets (such as GNP) not linked to some sort of equitable distribution of outputs (Thirlwall, 1989:8).

The development decades

Development as a concept has been most vigorously thought about during the post-war years with the declaration of the first United Nations 'Development Decade' of the 1950s. At that time and continuing virtually unchallenged into the second Development Decade of the 1960s, development was seen strictly in economic terms. Traditional structures of production were seen as incompatible with growth. For countries undergoing decolonisation (and other poor countries), this meant placing new emphasis on industrialisation with the deliberate (i.e. planned) consequence of shifting resources away from primary production (agriculture and rural activities) to the 'modern' manufacturing, industrial and to some extent, service sectors. The consequence of these post-war efforts, encouraged by former colonial powers and the burgeoning World Bank, was:

* rapid urbanisation;
* introduction of consumerism;
* neglect of rural development;[2] (and)
* the creation of economic (and thus social) dualism.[3]

These notions of *dualisms* are of fundamental importance in development studies. Cypher and Dietz are convinced that the industrialised capitalism imported by the colonising powers created a form of social and economic disintegration which has the potential to diminish the potential for national development. They go on to say that 'dual societies often exhibit profound urban bias' (1997:94–6). Todaro explains one aspect of dualism thus:

> The gap between the rich and the poor and between modern and traditional methods of production shows signs of growing even wider . . . it is the very growth of the stronger or 'superior' component of dualistic societies that keeps down, or at least is achieved at the expense of, the weaker or inferior element.
>
> (1982:94)

However, the idea of dualism within a nation, i.e. advanced industrial/economic enclaves and backward subsistence farming, can be criticised on the grounds that it assumes that no benefit leaks out of the advanced enclave and the rural areas remain traditional and poor.

Todaro's comments on the dualistic consequence of development are in part, a reflection of:

- measuring 'success' in development solely by means of the achievement of growth targets in GNP or GDP;[4] (and)
- an assumption of the inevitability of a disappearing peasantry.

As Toye remarks, 'it is important not to confuse economic growth, the expansion of the measured outputs of goods and services, with development' (1993:24). In discussing the social cost of capitalist growth in this dualism context, Isaak argued that:

> The priority of economic growth accentuates the distance between the winning and losing social groups, constantly redirecting resources to the winners in order to make them self-fulfilling prophecies and 'locomotives' for the economy. To the extent capitalist incentives are inevitable, so are manifestations of social inequality.
>
> (1995:241)

Isaak does not claim that such events are inescapable, but rather that 'the dual economy model can be a powerful mode of explaining growth-rate differences in . . . market economies' (1995:244).

The third decade of development saw the World Bank shift its official position from relative indifference about the distribution of the benefits of development, towards 'basic needs' focusing on a 'life sustenance' approach (Thirlwall, 1989:8; Toye, 1993:129). Goulet (1971) saw these basic needs as being one of three components of development, the other two being self-esteem (self respect and independence) and freedom (from 'want, ignorance and squalor') with respect to self determination. The advantage of material development, as Thirlwall (1989:8) reminds us, is 'that it expands the range of human choice'. The implication that this materialistic approach to development holds for Marxists (Peet, 1991; Pepper, 1993) and eco-centrists (Adams, 1990) is that development is simply equated with modernisation (technocentrism), taking no account of what is actually happening between regions and nations. Pepper, in his discussion on anthropocentric analysis of green political thought (i.e. where human-kind, rather than nature-at-large is the centre of attention) asserts that:

> Greens, anarchists and neo-Marxists of the new left unanimously see that 'modernisation' is really *dependent development* in disguise . . . in capitalist world 'development', underdeveloped nations are essential counterparts of the existence of developed ones. In other words, underdevelopment *results from* and is a vital feature of capitalism.
>
> (1993:26, italics in original)

What emerges from the above, is that while apologists for trickle-down economics[5] (Bauer, 1981) might claim that the absolute poor are better off than they were at the beginning of the colonialist era, it is distribution of income and wealth and not absolute measures of GNP that determine people's health, happiness and in effect, the degree of control they have over their own lives ('freedom' as Goulet termed it above).

Having described what is generally meant by, and some of the arguments pertaining to the notion of development, I now turn to the ways in which it has been defined.

Defining development

A good start on working towards or agreeing upon a definition of development might be found by looking at Ghatak's question 'Is per capita real income a valid index for measuring development of the LDCs?' (1995:34). Ghatak puts forward suggestions as to why the answer to this question might be 'no':

- without redistribution of income derived from economic growth, there can be growth but not necessarily development;
- unless the growth rate in economic outputs are capable of out-stripping populations growth, then here too, growth without development may occur; (and)
- 'dual' societies (where the gap between the very rich and the very poor is gravely significant) can be seen as evidence that economic growth for some without development for all has indeed taken place.

This means that definitions that imply 'growth' and 'development' to be the same thing or which assume development as a natural consequence of growth[6] should be rejected. In this sense, the following definitions by Balaam and Veseth, and also Todaro are found to be useful in that they acknowledge that benefits to civil society are an essential part of development. Balaam and Veseth in their book on the International Political Economy, define development as:

> the ability of a nation to produce economic wealth, which in turn transforms society from a subsistence or agricultural-based economy to one where most of society's wealth is derived from the production of manufactured goods and services.
>
> (1996:312)

In defining development economics, Todaro links the traditional concerns of economists (efficient use of scarce resources) with elements of political economy[7] with a view to setting up analytical mechanisms and theoretical perspectives that enable the study of:

> the processes necessary for the rapid structural and institutional transformations of entire societies in a manner that will most efficiently bring the fruits of economic progress to the broadest segments of their populations.
>
> (1982:501)

This of course implies a recognition of the need for government planning. While I started this section by suggesting that growth can take place without development, de Kadt makes it quite clear that development cannot take place without economic growth:

> Most Less Developed Countries cannot hope to create acceptable living conditions for the majority of their people without continuing economic growth and for many of them, especially the large number of smaller tropical mini- and micro-states, tourism represents one of the few apparently viable routes for such growth.
>
> (1992:75)

De Kadt's arguments are based around the involvement of a wide range of social and political institutions in the development process, thus while he does not specifically define development it may be inferred that his definition would be a holistic one in which 'the most formidable task on the road to sustainable development, and tourism development, is that of *building the institutions needed for policy implementation*' (de Kadt, 1992:73, italics added). This is a point of real importance because it raises issues of power and control, and, by implication, the need for appropriate scale, equitable distribution of benefits, long-term accounting, and (above all) the notion of the state supporting and encouraging regional and local initiatives. De Kadt includes NGOs and community participation citing Murphy's[8] (1985) view that tourism should be developed and managed as a local resource where local needs and priorities take precedent over the goals of the tourism industry.

Definitions of development should allow links to be made between economic growth, biological implications (health, life expectancy etc.) and individual welfare. The UNDP (1990) do this through their Human Development Index which is 'a new yardstick that provides a broad method by which inter-country and inter-temporal comparisons of living standards can be undertaken' (Ghatak, 1995:38). Figure 8.1 shows the four essential components of the human development paradigm which has moved on from the 'basic needs' model[9] of the earlier decade.

The Sustainable Human Development model used by the United Nations Development Programme in the 1990s is framed by the following tenets:

- people-centred: strengthening the capacity of civil society;
- poverty-elimination: transfer productive assets to the poorer sections of society;
- advancement of women and child development;

Empowerment: development must be **by** the people not only **for** them. *People must participate fully in the decisions and processes that shape their lives*

Productivity: people have the capacity to participate fully in paid enterprise. *Economic growth is thus a subset of human development*

through enhanced capabilities, the creativity and productivity of people must be increased so that they become effective agents of growth

Sustainability: access to opportunity assured for future generations. *Capital (environmental, physical, human) should be replenished*

Equity: access to equal opportunities with barriers to economic and political opportunities eliminated *enabling quality participation*

Figure 8.1: Sustainable Human Development model
Source: after UNDP, 1995: 12

- enhancing access to opportunity and reducing child labour;
- productive employment: including rural and informal sectors;
- protecting and regenerating the environment: sustainability;
- sustainable economic growth: renewable resources; (and)
- transfer of environmentally sound technologies.

While not exactly a definition of development it provides a logical way of monitoring whether development is having an effect on human welfare within a given country. Figure 8.2 shows some of the key thoughts in development which arise out of the tensions that exist between the 'economic enlargement' school of thought versus the Sustainable Human Development approach.

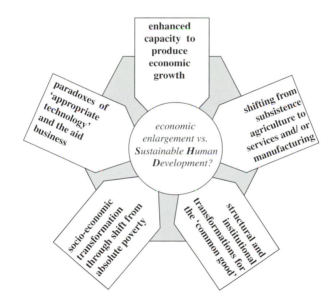

Figure 8.2: Some key thoughts in development

Conclusions and links to tourism and anthropology

In the broadest of terms, the development of development thinking has been subsumed by the two ideologies: Liberal and Marxist. This disjunctive was summarised by Isaak in the following way:

Whereas liberals perceive underdevelopment as a *condition* of countries that have not kept up with leaders in the world economy, dependency theorists view it as a *process* inherent in an asymmetrical [unbalanced] system that continually re-structures developing countries into underdeveloped positions. (1995:194)

In this view, which echoes the dualisms and the role of the 'new' WTO discussed earlier, Third World countries are locked out of 'normal' development by at least four factors:

• the entrenched power of international organisations such as the World Bank and the 'new' World Trade Organization, the procedures of which are designed to best suit the developed world (Harrison, 1993);
• a private sector banking system wherein exists 'an accelerated transfer of wealth from the poor countries to the rich' (George, 1989:5);
• the existence of multi-national corporations headquartered at centres within the developed world where countries vying for productive activity and investment from these corporations are thrown into competition with each other[10] (Balaam and Veseth, 1996:352); (and)
• the international transport system which is geared, for the most part, to the needs of the developed world (Britton, 1982).

These four factors have a very real resonance with tourism which is well summarised in Figure 8.3.

I started this chapter by noting Hobart's (1993) use of metaphors in the development debate and ended with Britton's (1982) asser-tion about global systems that are geared towards the needs of the developed world. The threads that draw these arguments and counter-arguments together, such as growth and distribution, industrialisation and primary products and the whole core–periphery controversy produce the two main strands to the debate, Liberalism and Marxism. It is these that provide the framework for analysing the links between tourism and development for LDCs.

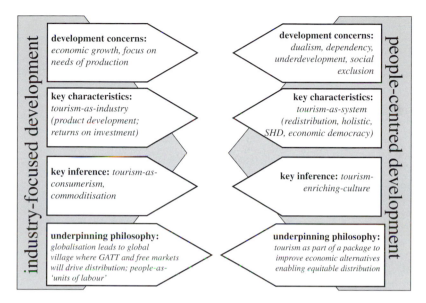

Figure 8.3:   Tensions and contradictions in tourism and development

Key ideas

- Too often development is what is 'done' to people (by a variety of agents including the UNDP and World Bank) rather than a reflection of what they want or need;

- the emergence of dual economies is a result of colonialism and has caused continuing structural problems in Third World countries;

- development can be defined from both an economic growth and human development perspectives: both have their pros and cons, but increasingly per capita GNP or GDP is found to be unsatisfactory; (and)

- the whole development debate is framed by two opposing ideologies, capitalism (or liberalism) versus Marxism which has caused a continuing and unresolved tension about how countries best 'develop'.

Questions

1   What are the economic advantages and disadvantages of accepting inward investment in tourism for a developing country?

2   What is the justification for a trickle-down economic development policy?

3   How can the need for local, community-based development be accommodated within national development planning?

4   How can tourism be best used as a tool for economic and social development in a developing country?

5   Are there ways in which certain types of tourism systems (identified in Chapter 2) should be linked to development aims and objectives?

6   Are there ways in which different types of tourists (as identified in Chapter 3) can have different developmental impacts?

7   In what ways might an anthropologist help create a middle ground for those with divergent interests in community welfare, tourist development and national development upon which to meet?

Key readings

It is difficult to separate this subject from political economy. This being the case, the first books to read are Colin Hall's *Tourism and Politics* (1994, Wiley) and Linda Richter's *The Politics of Tourism in Asia* (1989, University of Hawaii Press) and Bryan Farrell's *Hawaii, the Legend That Sells* (1982, University of Hawaii press). Then there is John Lea's somewhat dated but still useful *Tourism and Development in the Third World* (1988, Routledge) and (if you can get hold of it) John Bryden's *Tourism and Development: a Case Study of the Commonwealth Caribbean* (1973, Cambridge University Press).

In a more general sense, Michael Todaro's substantial book (brought right up to date with the 6th edition) *Economic Development* (1997, Longman) will allow you to explore all the basic theories of development and lead you into further reading (in a sense prepare you for it). In addition, there are some specialist books that are useful such as James Cypher and James Dietz's *The Process of Economic Development* (1997, Routledge) which, while very technical, enables insights to be developed and creates an awareness of the main arguments.

SPECIAL ANNEX TO CHAPTER 8: THEORETICAL PERSPECTIVES ON DEVELOPMENT

Evolution of development

Meek (1976) suggests that the first theory of development (in a context that would be recognised today) emerged from the mercantilist, burgeoning industrial period towards the end of the eighteenth century: the Enlightenment. This was a period of optimism, with a potential to shape society for the common weal. As darkness and ignorance gave way to light and the rational mind, it was coupled with the widespread conviction that, with scientific knowledge, society could be liberated and reconstructed thus moving it away from the antiquated ruling groups. It was however, also recognised, even at that early stage, that capitalism not only brought about wealth, but created cultural, social and economic problems (Court, 1967).

Framed by this ideology of rational, 'scientific-being' willing to be democratically governed, Meek characterised the Enlightenment model of development as being arrived at by progressing through four stages:

1   nomadic hunting and gathering clans;
2   pastoralism and settlements in hamlets and nascent villages;
3   settled agriculture with an understanding of formal land usage arrangements;
4   nascent capitalism and 'commerce' including money exchange.

This last stage was seen to be the steady-state, whereby all of society would benefit from commercial activity. The model did not last beyond the beginning of the nineteenth century. Internal logic problems, centring around how society passed from one stage to the next (the transition stages) and the paradox of different modes of subsistence existing in parallel, ensured its decline.

During the nineteenth century, four distinct models of development emerged from the Enlightenment period:

1   *evolutionary development theory* which was based around the environmental determinism of Darwin whereby species evolve over time to become increasingly efficient. This can be seen as a harbinger of the *laissez faire* model;

2   *technocratic development theory* which drew upon the scientific, rational thinking posited through the Enlightenment model to introduce the idea of planning for industrial capitalism. The technocratic belief was that society was advanced by the development of ideas. Education was high on their agenda and the development of a 'new morality' (i.e. no greedy capitalists and no lazy workers) would resolve class conflict;

3   *socialist development theory* (Marxism), was powered by concern over the growing contradictions arising out of industrialisation and capitalism. For a Marxist (of whatever era) the dominant paradox remains the unresolved relationship between the forces of production (i.e. the way in which nature is transformed by labour) and the mode of production (i.e. the system of political economy);

4   *populist development* theory, where emphasis was placed on the benefits of development for the common people, especially rural populations.

These first three models could, in their own ways, be seen as underpinning modernisation or 'developmentalism' as Adams (1990) describes it. I include Marxism here because of its technocratic, Eurocentric nature. Adams continues his argument on modernisation, 'this dominant development paradigm embraces both capitalist and socialist approaches to development'. Citing Friberg and Hettne (1985) Adams goes on to identify a paradox:

> the capitalist societies of the West and the state socialist societies of the East are two varieties of a common corporate culture based on the values of competitive individualism, rationality, growth, efficiency, specialization, centralization and big scale.
>
> (Friberg and Hettne, 1985:231 cited by Adams, 1990:70).

Development in these three models is unilinear, cumulative and predetermined through a set number of development or societal stages inducing, as Waters asserts, 'value shifts in the direction of individualization, universalism, secularity and rationalization' (1995:13).

The main problems about the populist approach were:

- there is no guarantee that the exchange of goods for money would reflect the labour content;
- the impact of the international market was difficult to assess; (and)
- the seemingly intractable problem of ineffective government administrative support.

Tanzania's experiment with agricultural collectives, the *Ujamaa* villages, is probably the most well known example of populist socialism (Balaam and Veseth, 1996: 323). However, this experiment in 'forced collectivisation' (Bauer, 1981) failed to meet basic needs. The fundamental flaw was that by concentrating on an export strategy for rural products (while at the same time importing meat and milk products) it did not address a major paradox of development as Third World countries become exposed to global capitalism: the dynamic tri-vergence of domestic resources and capability; domestic demand distorted through raised expectations; and the basic needs of the masses (Peet, 1991:166).

Unlike the first three, this model is no progenitor of, nor does it owe allegiance to, modernisation theory. In this sense (even taking account of the three problems outlined above) populist development theory remains enveloped by the dependency framework (explained elsewhere in this chapter), in that enhanced agricultural production requires capital and skill inputs which inevitably (either through trade or aid) entail contact with global capitalism, especially since the collapse of the 'Second' World.

Having described some of the main aspects of development and traced the development of development thinking, I now turn to the work of Rostow, which has been described as an integral part of the United States' cold war containment of communism (Mehmet, 1995) and as such was a powerful influence on Western aid to the Third World for at least the three decades following the Second World War. This brief examination of Rostow will help set the stage for its ideological Nemesis, dependency theory, that follows immediately afterwards.

The work of Rostow

Rostow (1960) proposed that it is possible to categorise all states as lying along a continuum of five stages of development. The first of these is *traditional society* which is generalised by Rostow and his followers as having primitive technology, a spiritual/fatalistic relationship with the physical world, and hierarchical/hereditary social structures. Rostow saw these conditions of being in the way of growth: traditional culture, attitudes and values as a barrier to growth.

The second stage is termed the *pre-conditions for take-off*. These conditions include the acknowledgement and use of modern science for production, international trading, traditional ways of production replaced by 'rational management' and the notion of the good of capitalism. Rostow's third stage is termed 'take off'. In the case of the first countries to 'take off' (notably Britain in the nineteenth century) the drive was both technological (in the sense of new inventions and the mechanisation of production and division of labour) and political (in the sense of imperialism). Rostow is very specific about defining the investment aspect of 'take off'. He stated quite clearly that during the 'take off' stage investment must rise to 10 per cent of national income (Rostow, 1960:6–31) with the idea that this reinvestment will increase paid labour and create an increasing number of entrepreneurs. This idea about levels of investment is important as it became one of the foundations of post-war aid and economic assistance, and remains, to some extent, one of the underpinning philosophies of the World Bank and the International Bank for Reconstruction and Development (IBRD). With these conditions in place, the nation is propelled into self-sustaining growth, and the two final stages: *the drive to maturity* and the *age of high mass consumption*. Rostow was adamant that:

> These stages are not merely descriptive. They are not merely a way of generalizing certain factual observations about the sequence of development in modern societies. They have an inner logic and continuity . . . they constitute . . . both a theory about economic growth and a more general, if still highly partial, theory about modern history as a whole.
>
> (1960:12)

The basic assumption for Rostow and his supporters was very simple: the more a country could put into national savings which in turn was to be invested in economic development, the faster that country could grow. The gap between the ability (as opposed to the propensity) of a country to achieve certain domestic savings targets could be made up with transfers of capital from the industrial countries (i.e. aid from the West), 'the lesson [of economic history] . . . is that tricks of growth are not that difficult' (1960:166). In this sense, Rostow was not a pure *laissez faire* capitalist but rather, with his support for government planning and intervention, a state capitalist. This tenet was a reflection of the success of the *Marshall Plan* that re-constructed Europe after the Second World War.[11] So, while some continue to place value on Rostow's work (e.g. Thirlwall, 1989) for its ability to provide insights into the development process, the deterministic view that development will come about through the removal of obstacles, the right mixture of science and technology, the emergence of new élites willing to fabricate industrial society, and the development of social conditions that encourage entrepreneurial risk-responses to material incentives, does not take account of the complexities of the global political economy within which Third World countries are enmeshed.

Hoogvelt in discussing Rostow's later work (i.e. that produced in 1978) takes these arguments further by highlighting Rostow's hegemonic tendency:

Rostow is all for a stable and effective partnership between the North and the South and for a 'common agenda' for the 'world community' provided this partnership and this community is organised and directed under the global leadership of the USA. In fact, Rostow quite openly pleads for a return to the USA's world hegemonic role – which he observes has declined in the 1950s and 1960s . . . [because] . . . the inputs to sustain industrial civilisation must be expanded.

(Hoogvelt, 1982:137)

Hoogvelt's mention of 'common agendas' and 'world community' are deliberately ironic references to the New International Economic Order (NIEO), first mooted in the early 1970s as a way of combating the dominant world trading paradigm, and which has thus far failed to make any impression (Balaam and Veseth, 1996:317–318).

Development models in the context of the Third World have been overshadowed first, by the incidental racism of Darwinian theory, and second, by the Eurocentricity of both Rostow and Marx, as illustrated by Mehmet who is critical of both of them, 'Rostow, the economic historian was like Marx, his ideological Nemesis, Eurocentric; their theories were linear, deterministic and guided by Western history *perceived as the universal law of development*' (1995:68, italics added).

What Rostow and later, Bauer decline to discuss is the inherent hegemony of the West (more particularly, the United States) in the global economy. It can be inferred that this is because they both believe, as conservatives, in a 'natural order' of things, the 'things' in this case being trading positions and power. The corollary of this global power structure is *dependence* where the international and domestic economic priorities and policies of developed countries frame the economic development of Third World countries.

These conflicts and paradoxes fuelled the development of dependency theory, to which I now turn.

Dependency theory and beyond

Dependency theory seeks to examine 'the effects of imperialism on overseas territories in an attempt to explain the roots of backwardness' (Hoogvelt, 1982:165). The basic assumption for dependency theory is, as I have mentioned above, the reliance of Third World countries upon the economic policies of the developed countries. A further corollary asserted by dependency theorists is that of *underdevelopment*, whereby dependency results in persistent low levels of living as dependent economies are distorted towards the needs and predilections of the metropolitan centres; growth for the underdeveloped economies is 'a reflection of the dominant countries' (Peet, 1991:45).

For dependency theorists, development and underdevelopment are two sides of the same coin: surpluses from the exploited countries, generated first through mercantilism and later through colonialism, had the combined effect of developing the metropolitan countries and under-developing the peripheral countries. The dependent situation became heightened with the development of local élites, 'whose economic interests became . . . intertwined with . . . the advanced capitalist states, and whose cultural lifestyles and tastes were a faithful imitation of the same' (Hoogvelt, 1982:166).

Two writers featured heavily in the development of dependency thinking. The anti-imperialist work of Raúl Prebisch, a former head of the Central Bank of Argentina, one time director of the United Nations Economic Commission for Latin America (ECLA) and Secretary-General of the 1974 United Nations Conference on Trade and Development (UNCTAD) conference which established the NIEO, and the neo-Marxist, André Gunder Frank (Peet, 1991:45) who reworked and furthered Baran's (1967 in English, 1957 in Spanish) work on the umbilical links between dependency and underdevelopment.

Prebisch was one of the first economists to 'question the mutual profitability of the international division of labour for developing countries on existing lines' (Thirlwall, 1989:370). Prebisch attributed Latin America's underdevelopment to the structural characteristics of the global system, arguing that 'Latin America's peripheral position and primary exports were the causes of its lack of progress, specifically because of a long-term decline in the terms of trade of the periphery' (Peet, 1991:44). These are the conditions under which, according to Waters, development, becomes 'impossible':

> Rich states have dynamic economies committed to technological advancement in which monopoly corporations and effective labour unions can hold up the prices of manufactured goods. Meanwhile poor states have feeble investment patterns and a disorganized labour force . . . This produces a consistent tendency towards increasing disparity between the prices of manufactured goods and raw materials that makes development impossible.
>
> (1995:109)

Through this argument, Prebisch suggested that the solution to underdevelopment was through modernisation, that is to say, investment in industrialisation. The consequence of this approach was a move away from primary (agricultural) production. The main support structure for this programme of industrial investment was import substitution through internal growth, and the development of 'infant industries'[12] sustained by fiscal and non-fiscal tariffs (i.e. import taxes and import restrictions/quotas).

Import substitution failed as a long term policy, becoming a remedy 'seen as a cause of the economic illness' (Peet, 1991:45). Wages did not rise to stimulate domestic demand, balance of payments worsened and income distribution became less equitable (Hoogvelt, 1982:168). As Todaro (1982:314) explains, four undesirable outcomes arose from it:

1   foreign firms taking advantage of mobility of production and excessive tax breaks were a major, yet unintended, beneficiary;
2   governments often subsidised importation of the heavy plant and equipment necessary to set up factories;

3   changes to official foreign currency exchange rates designed
    to help industrialisation through cheap importation of capital
    further encouraged capital-intensive production, thus inadver-
    tently disadvantaging the export value of primary goods which
    (because of the exchange rate) become expensive on world
    markets; (and)
4   the intention of forward and backward linkages (i.e. industries
    buying from and selling on various components needed in the
    production process from other domestic firms) becomes
    distorted by the inefficiencies inherent in a non-competitive
    domestic market.

Frank (1966) was critical of Prebisch's modernisation perspective where Latin America's subservient position relative to the
United States and the industrialised world was caused by unequal
trade, rooted in an international division of labour wherein Third
World countries were destined to be the producers and exporters of
commodities, the prices of which were set by commodity markets
in those same developed countries. For Frank, 'attributing underdevelopment to traditionalism (or feudalism) rather than capitalism
is an historical and political mistake' (Peet, 1991:46). Frank's most
tangible contribution to the dependency debate is his bringing
to the forefront of the development discourse, the concept of the
world being polarised into interdependent 'metropolis centres'
(usually meaning the industrialised nations of the North) and
'satellite peripheries' (former colonies and other primary producing
countries of the South).

Hoogvelt (1982) claims this dependency framework, and Frank
in particular, continuously misses the point; the issue of whether
or not élites deliberately act as tools for the new imperialists is a
red herring. At a deeper level, what remains as the continuing
paradoxes of dependency thinking is its focus on Third World
countries as being locked into pessimism (Kay, 1989): a 'negative
self-fulfilling prophesy' as Isaak (1995:195) describes it, and the
unresolved issue of 'strident nationalism' (Toye, 1993) as a
response to the disadvantaged position of Third World countries in
the International Political Economy.

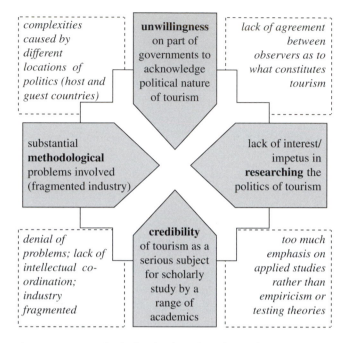

Figure 8.4: Factors in the lack of political analysis of tourism
Source: after Hall, 1994: 4

Finally, in order to link this special annex back to our over-riding theme of tourism through the unifying thread of politics, we look at Michael Hall's reflections on why tourism lacks the necessary political analysis (Figure 8.4).

Notes

1 Anthropology

1 It is worth noting that the Chinese were conducting recognisable anthropological studies in the thirteenth century (Murray, 1994, *Anthropology Today* 910) and Hodgson, M. T. (*Early Anthropology in the 16th and 17th centuries*, University of Pennsylvania Press) reports on attempts at gathering data on what today would be called tribal peoples in the sixteenth and seventeenth centuries. The Moroccan scholar ibn-Khaldun wrote up what were, essentially, ethnographies of the people he encountered in his travels.

2 The Enlightenment was a seventeenth century literary and philosophical movement against superstition, ignorance, 'traditional' knowledge and accepted wisdom. It was framed by increasingly critical analysis of the Bible, the execution of Charles I which swept away the idea of a Divine Ruler and an increasing belief in science as a way to explore and further knowledge. Key figures were John Locke (1632–1704), René Descartes (1596–1650), Goethe (1749–1832), Benjamin Franklin (1706–90). Key themes and consequences were a belief in Newtonian physics (the idea that the universe was governed by discoverable laws) and, in effect, the French Revolution. Adam Smith's *The Wealth of Nations* (1776) also typifies the Enlightenment era as it sought scientific, rational solutions to economic problems.

3 Jean-Jacques Rousseau's idea of a 'man of nature' or 'noble savage' as a primitive unspoilt by modernity has acted as a symbol for 'romantic conservatism which seeks a sight of cohesion in the distant present' (Hiller, 1991:65) and as a sort of antidote to industrial life.

4 Classically, this must exceed one calendar year so as to encompass all the annual events pertaining to the culture and society under observation.

5 Participant observation is another name for, or way of describing ethnology, the close and long-term study of a particular society. It is taken for granted that the anthropologist conducting the study will first learn the appropriate language, so that the observations are taken at first hand and not through the barrier of translation.

6 I attempt to describe and define this a number of times in this book. It is a method of studying a thing by way of analysing the structures that comprise it (so a structuralist study of public transport would be about why and how various passenger groups use buses and trains, trade unions, employees, managers etc. rather than routes and time-tables).

7 Malinowski first categorised a system of symbolic dualisms during his time in the Trobriand Islands. Some of these were: 'below : above', 'moon : sun', 'spirits : men', 'invisible : visible', 'immortal : mortal'. Lévi-Strauss re-examined Malinowski's work and added further, even more complex dualisms: 'peripheral : central', 'cooked : raw', 'profane: sacred' and 'marriage : non-marriage'.

8 The frequent use of the male gender to explain is a reflection of the uninformed time in which this was written. It should not detract from the value of what these authors are saying.

9 This provides fertile ground for philosophers and other academics to debate the nature of truth, it cannot be something to explore in this book.

10 There is another view of myths in the context of history. Amon Saba Saakana asserts that the two were brought together in the age of Imperialism and the Enlightenment as a form of narrative discourse infinitely superior to the myths of the 'untamed, unchristian savage'.

11 Applied anthropology is a term defined by Howard (1996:400) 'as research and activities intended to produce a desired socio-cultural condition that optimally will improve the lives of people concerned'. However, he goes on to say that it was first coined by Lane Fox Pitt-Rivers in 1881 and as a sub-discipline, was used to train colonial officers and 'facilitate the establishment of colonial authority'.

12 The term 'Third World' has its roots in the cold war division of the globe into a First World (the developed West), the Second World (countries within the Soviet empire: the Eastern Bloc) and all the rest: the Third World. It is now seen as somewhat outdated, but a number of scholars still use the term which is seen by them as no worse or patronising than many other euphemisms that are used to describe poverty and underdevelopment.

13 This is where anthropologists step aside from their 'cold' scientific examination of society and actively become engaged in the 'hot' business of active campaigning and politics on behalf of the people they are studying: this is especially so in the case of indigenous, tribal peoples.

14 Malinowski was the most well known proponent of the functionalist approach which proposed that each part of culture functions as a whole to fulfil basic needs (food, shelter, defence and reproduction) and derived needs (laws and economic systems) in order to assure biological survival. Neofunctionalism was an attempt to recognise conflict as being important to culture (colonialists tended to dampen

or hide conflicts as it tended to reflect unfavourably on their rule). Gluckman's point was that societies are just as likely to form alliances and unite against something as for it.

2 Tourism

1 This is a very troublesome term! One way to explain it is to start by describing 'modernism' or 'modernity' which began in the sixteenth century when the first clashes between 'Ancients' and 'Moderns' (i.e. discovery and conflict between societies with manufactured technology and those without) resulted from voyages of discovery, through the enlightenment up until December 1910 (the moment when, according to Virginia Woolf, human nature changed!). This is less of a joke than it might at first, seem. The mass-producing industrialised world was seen as 'thoroughly modern' with explanations about how it worked and break-throughs in science coming thick and fast. It took the rise of 'common-cultures' (made possible by mass communication systems and growing democracy), the art movements (starting with Fauvism, through post-impressionism, Dadaism, surrealism, etc.), the overt politicisation of art and the critical deconstructionist approach to literature and architecture in which all was analysed for hidden meanings to 'break' the modern world and enter the post-modern era. (A word of warning, post-modernism is by no means accepted as useful by all cultural observers and commentators.) The term came to be widely used in the 1970s to describe the rejection by some architects of the so-called Modernist Movement in architecture (i.e. the use of concrete in an austere, almost brutalist way). Post-modernists attack all systems of thought and meaning from traditional religion to scientific reason, anything with claims to 'an objective truth'. Post-modernist hold that the constant flow of media images and bombardment of information means that media defines reality with a constant merging of fact and fiction.

2 One problem arising, is the blind acceptance of the figures promoted by the WTTC concerning the size of the tourism industry. This from a recent World Bank newsletter (dated 19 June 1998: 'Studies by the WTTC indicate that tourism is the world's largest generator of jobs. It accounts for 10 per cent of total employment in 1997 and will create jobs for an estimated 230 million people in 1998. By 2010, this figure is expected to reach 283 million. The travel and tourism industry accounts for $4.4 trillion of economic activity world-wide.' What is disturbing here is that these figures are inflated by every conceivable tourism activity. So students and other observers should know that the entire civil aviation industry world-wide is included, passenger trains, opticians who supply eyecare to tourism workers, those who work in the suntan industry etc. To date, I have seen no sustained

effort to challenge the WTTC on its statistics, they have become part of the myth that surrounds tourism.

3 The phrase 'The Tourist Gaze' is Urry's own. He published his book with the same title in 1990. His central idea is that 'there are systematic ways of "seeing"' what we as tourists look at, and that these ways of seeing can be described and explained. He also suggests that a host of other social practices such as shopping, sport, culture, hobbies and tourism have increasingly blurred divisions.

4 A fairly neat solution to this developed over time in the British seaside resorts of Great Yarmouth and Gorleston. Casual workers would be employed in the summer season in the tourism industry and then as the harvest and fishing season arrived in autumn and winter they would work in the nearby Bird's Eye frozen food factory! This 'system' lasted no longer than two decades as Bird's Eye rationalised its operations and closed the factories in East Anglia.

5 In a sense, I thank my own children for this observation. Their holidays seem to consist of a mass tourism product of sand, sun and sea taken at places like Ibiza and Skiathos. They enjoy their two weeks of fun in the sun and demonstrate no visible signs of alienation!

6 Some are troubled by this term, for instance MacCannell (1992) asks 'What can be said with precision about those aspects of personal identity that are marked 'ethnic'? . . . In the social science literature, the term 'ethnic' occupies the space between biological and genetic conceptions of race, and anthropological theories of culture. Ethnicity has never been subject to rigorous definition.'

3 Tourists

1 I am being a little cynical here, but there is a perception of what the word 'tourist' means to the general public, and what it means to official organisation in whose interest it is to boost tourist arrivals numbers. So, in one particular country where I was conducting research, visitor arrivals figures always hovered around the million mark, this was the figure that was trumpeted as being the tourist numbers. Closer examination of the statistics revealed that three-quarters of these arrivals were short-term contract workers from neighbouring countries. The whole issue of tourist statistics is riddled with inconsistencies and lack of international agreed definitions.

2 What I tell my students is how, as a young and relatively poor family, we would take a day out to Brighton (a seaside resort about 30 miles south of London, England). We would buy our bread, cheese, snacks and drinks from our local Safeway supermarket, make up a picnic, fill up with petrol at our local filling station, drive to Brighton, park outside somebody's house (thus avoiding the town's car parking fees), eat our picnic, have a swim in the sea and walk on the cliffs, put our rubbish in their bins, pee in their free public toilets, and go home. The

3

effect of staying over would, of course, dramatically have changed our economic impact on Brighton.

3 In this sense, novelty means visits by tourists are infrequent enough for local populations to be interested in seeing or speaking with (or even giving a bed to) passing strangers, the tourists are still a novelty rather than an everyday occurrence.

4 Philip Pearce makes the important point that individuals might undertake some sort of 'travel career ladder' whereby tourists develop a range of motivations as they 'learn' to travel and develop the capacity to experience it at a number of levels (Pearce, P.L. (1988) *The Ulysses Factor: Evaluating Visitors in Tourist Settings*, New York: Springer Verlag). If this book is difficult to trace, a critical analysis of Pearce's work may be found in Ryan, C. (ed.) (1997) *The Tourist Experience: a New Introduction*, London: Cassell.

5 In a letter dated 24 December 1996 to Andrew Holden, Stanley Plog stated that he had moved a little beyond this classification and now tends to group traveller types under the more user-friendly headings 'venturers' (formerly the allocentrics) and 'dependables' (the former psychocentrics).

6 The term 'otherness' here is an important one. Edward Said, in his classic work *Orientalism: Western Concepts of the Orient* (1978) uses the word 'Other' (note the capital 'O') to emphasise that descriptions of foreigners, especially those from 'exotic' places are social constructs to reinforce visions of European cultural and intellectual superiority. Thus 'Other' means the opposite of 'us'. The word 'neutralising' in this quote is used in the sense that marketing and promotion of destinations has the effect of creating a bland image so as not to offend any potential market segment. This propensity for making destinations 'all the same' is called 'homogenisation' by its critics (Burns and Cleverdon, 1995).

4 Culture

1 This can be both in the sense of visitors wanting to experience new forms of culture or the social impact of tourism on the culture(s) of a destination.

2 In some cultures, such as that to be found in Bali, Indonesia, 'art' is such an integrated part of daily 'ordinary' life that no specific word exists for it.

3 I am grateful to professor Tom Selwyn for these ideas on differences between 'race' and 'culture'.

4 Another example is the continuing popularity and high price of 'trainers' (sports shoes) such as Nike. This remains a constant source of fascination for those (such as myself) who are outside observers of the youthful 'society' where such expensive trainers ascribe power and legitimacy to their owners!.

5 The term 'commoditisation' or 'commodification' ('commoditization' in the US) is an important one which gets further space in this book. In brief, it is the process by which some object or other 'thing' is ascribed value by the price put upon it. To cite Nash (1996:14) in commenting about Greenwood's work: 'The concept of commoditization was used to describe the tourism-induced transformation of the meaning of [a festival] . . . into spurious superficialities associated with market exchange.' It may well have lost much of its relevance now with the growing sophistication of tourism analysis and the forces of globalisation.

6 This is a reference to a style of industrial management associated with Henry Ford and mass production. It implies division of labour and heavy-handed, top-down management. See also 'scientific management' and the work of Bernard Doray (1988) who describes Henry Ford as turning workers into appendages of machines. His current concern is that such methods are now being applied to clerical and intellectual labour in a fit of what he thinks of as 'rational madness'.

7 Shaw and Williams (1997:3) say this about Butlin's: 'Their advertising slogan for the Clacton camp [was]: "Holidays with pay – Holidays with play. A week's holiday for a week's wage."'

8 'Cannibal Tours' is a most amazing ethnographic study of tourists, offering a series of ironic vignettes against a backdrop of Papuan mystery: the archetypal Other. The continual theme that threads the film together is that of mutual misunderstanding. It is essential viewing for any student of the anthropology of tourism. The London-based NGO, Tourism Concern has a copy it sometimes lends out.

5 The anthropology of tourism

1 The essay, 'Tourism as an Anthropological Subject', was published in one of the most prestigious American journals, *Current Anthropology* (Vol. 22 No. 5 October 1981). It had the intention of critically assessing 'the anthropologists' thinking about this fascinating subject'. The essay attracted several critical responses and these are well worth reading.

2 Magnum Images, a cooperative photographic agency, has published a collection of photographic images under the title *Ritual* (1990, London, André Deutsch) which it describes as ranging 'from the religious to the profane, from holy sacraments to popular celebrations, carnivals and pilgrimages, parades and political conventions, fair-rounds and children at play'. The images shown include a Barmitzvah in New York, circumcision among the Masai, a Bavarian beer festival, a Japanese tea ceremony and so on.

3 Incidentally, think of degree awarding ceremonies at universities: they follow these characteristics very closely.

4 The traditional anthropological description of totems/totemism is first

of all based on societies (such as hunter/gatherer types) who live *in* nature rather than *against* nature. Second, this type of society often has spiritual rules that frame supernatural relationships and unity with plants and animals. Ritual gatherings to worship the particular items were part of social cohesion and solidarity.

5 Even though in so-called post-modern societies, ritual is usually secular as opposed to being 'sacred' it must confirm whatever beliefs are crucially important and of emotional impact (e.g. The Doors' singer Jim Morrison's tomb in the Pére-Lachaise cemetery in Paris, or the Golden Arches of McDonald's).

6 There is a connection here between this and Jafari's interpretation of core–periphery relationships between tourism generating and receiving areas which is illustrated in this book as Figure 5.3. However, Jafari does not go so far as to call it imperialism.

7 Semiology and semiotics are more or less the same thing. Saussure coined the first (and which is thus used by Europeans), Pierce the second (and is used by Americans). The whole subject is a linguistic/cultural minefield that I am reluctant to enter. MacCannell defines a tourist attraction as 'an empirical relationship between a *tourist*, a *sight* and a *marker* (a piece of information about a sight) . . . the marker may take many forms: guidebooks, information tablets, slide shows, travelogues, souvenir matchbooks, etc.' (1976:41). For a clearer picture of semiotics and tourism it is worth reading MacCannell's Chapter 2 on 'Sightseeing and Social Structure'.

8 Think of phrases like 'This is the spot where . . . (etc.) . . . took place' as the tour guide points towards a particular building or piece of pavement. The 'Grassy Knoll' as discussed in relation to President Kennedy's assassination is one such example. There are plenty of others.

9 Curiously, it is only the last category that anthropologists have really got to grips with. It is Barthes in general, and MacCannell and Urry in particular that still hold the ground on semiotics and tourism, while Britton's seminal article (1982) on the political economy of tourism remains thus far unchallenged, although Raoul Bianchi has investigated (though not yet published) this through his PhD field work.

10 Boorstin says: 'These attractions offer an elaborately contrived indirect experience, an artificial product to be consumed in the very places where the real thing is as free as air. They are ways for the traveler to remain out of contact with foreign peoples in the very act of "sight-seeing" them. They keep the natives in quarantine while the tourists in air-conditioned comfort view them through a picture window. They are the cultural mirages now found at tourist oases everywhere' (1964:114, cited in MacCannell, 1976). The point that MacCannell makes about this is that Boorstin claims that the market demands by the movement of mass tourists cause the tourist industry

to respond in ways that always present a 'staged' or inauthentic, sanitised, cleaned up event or scene.

11 In this case Feifer is discussing the 'post-tourist' 'who almost delights in the inauthenticity of the normal tourist experience' (Urry, 1990:11).

12 Note the ironic reference to Rousseau here.

13 Maslow is renowned for his 'Hierarchy of Needs' theory, which puts forward the idea that in life we are all driven by a hierarchical set of needs, starting with the basic need for food and shelter, approval from peers etc. into the final stage: 'self-actualisation' when life's jigsaw comes together to provide a balanced state of well-being. Although initially attractive because of its simplicity, it is a clearly a culture-specific theory framed by capitalism and the American Way. It does not transfer to societies that value community and others above self.

6   Issues in the anthropology of tourism

1 Although of course, this is precisely Graburn's point. This fundamental contrast between 'home' and 'away' is a major motivating factor in leisure tourism. Doing 'nothing' away on holiday is acceptable, doing 'nothing' at home is (of course!) vaguely suspect. David Brown (1996:36) points out that Graburn's analysis misses those who are in some sort of enforced leisure at home, such as the unemployed.

2 In a sense, we are back to our basic definitional dilemma. A pilgrim travels for a reason which is not related to work, therefore he is a tourist, but the sets of arguments presented in this chapter go deeper, we are not interested in definitions as part of some manic categorisation fetish, but in the pursuit of meaning and causality.

3 Cohen makes the point that some forms of tourism, such as package holidays, are 'institutionalised' through the normal commercial actions of tour operators. Charter flights, ground transfers, excursions etc. are all taken care of. The tourist exists in an environmental and psychological bubble protected from anything that might go wrong at the destination. Independent tourists who make their own arrangements and travel by local transport etc. have no such institutional setting surrounding their touristic activities.

4 If the controlling-institutional setting is strong enough (and in mass tourism destinations tour operators work hard at directing and controlling tourists so as to maximise revenue generating opportunities) then apart from the artificial context of servant–master relationship, such meetings are not inevitable, or if they do take place they are under highly controlled conditions.

5 Where all aspects of life are changed by the sheer pace of transformation is driven by hyper-competitiveness.

6 The Virgin Holidays brochure which features California, Hawaii and other popular US destinations.

7 This was a form of racism. There was an assumption that the collective past remembering of tribal peoples were inferior to the recollections and written history of colonising powers. This is strange, at school (Chiswick, West London, 1960–65!), I can remember being told about Robert the Bruce and a spider, King Harold and the arrow in his eye, King Alfred and the burnt cakes and other such stories as historical facts! I do not recall being taught anything (in primary or secondary school) about more interesting historical 'facts' such as Cromwell's activities in Ireland.

8 In my teaching, I constantly warn students of the dangers of oversimplifying tourism by calling it an industry. It does not and cannot (because of its diverse nature) exhibit the characteristics of an industry (such as homogeneity, and clear sectoral boundaries). I prefer the term economic *sector*. This is an important point, but not one that should cause sleepless nights.

7 Globalisation, tourism and hospitality

1 Flags of Convenience where ships are registered in Panama or Liberia or other countries where regulations, unionisation and taxation are 'liberal'. This has led, for example to the collapse of the British merchant fleet which now barely exists.

2 Modernisation is a contentious term in that some reject its pre-dilection for concentrating on that which is quantifiable such as employment, literacy, GDP while neglecting 'quality of life factors such as happiness' (cf. Robertson 1992:11).

3 Historic political events arising from *glasnost* and *perestroika* have accelerated the shift away from the familiar post-war East–West divide ushering in a political economy characterised by the tripolarity of a US dominated Americas, a Japan dominated South-East Asia and German dominated western Europe.

4 GATT, the General Agreement on Tariffs and Trade, came out of the Bretton Woods conference in 1945. It was designed to intervene in international trading disputes and kick start the immediate post-war global economy. The World Trade Organization (in this book referred to as the 'new' WTO so as to avoid confusion with the World Tourism Organization) is a more formal and more powerful body.

5 Cynthia Enloe's book *Bananas, Beaches and Bases: Making Feminist Sense of International Politics* (1989) that in asking the seemingly simple question 'where are the women?' engages in a radical and fascinating exploration of the international politics of masculinity. Her chapter 'Carmen Miranda on my Mind: International Politics of the Banana' reveals a complex set of pressures on growers and workers.

8   Charting development thinking

1   The term 'holistic' is quite central to my thoughts of tourism development. Holism means approaching subjects as systems, rather than studying their component parts in isolation from each other. A systems approach infers that phenomena are more than the sum of their parts, a fitting starting point for the study of tourism and development.

2   Except for the controversial introduction of chemically-dependent, capital-intensive methods (cf. van der Ploeg's *Potatoes and Knowledge* in Hobart, 1993; Adams, 1990:7).

3   In agriculture, dualism is characterised by 'large numbers of peasant cultivators who control extremely small parcels of land, while a few large land-holders, who constitute the landed oligarchy, own and vie with agribusiness transnationals to control vast quantities of land' (Cypher and Deitz, 1997:337). In manufacturing it is where enclaves of capital-intensive modern industries producing sophisticated goods for export exist side-by-side with traditional, low-technology modes of production oriented to unsophisticated and undemanding domestic markets.

4   During the first two Development Decades, measuring development by economic means was the favoured method by the World Bank and other agencies involved in development. However, it takes no account of informal or non-monetary exchange, and carries an urban bias: it measures what is important to capitalism.

5   'Trickle-down' approach to development is based on the idea that improvement to a nation's general economic condition (as measured by GNP) would automatically 'trickle-down' to the mass of population in the form of employment. It emphasises growth at all costs, poverty alleviation and equitable distribution of income will sort themselves out. Reagan and Thatcher were proponents of this for their own countries and for others.

6   Cf. Bauer, 1981 especially his Chapter 11, 'Broadcasting the Liberal Death Wish'.

7   Political Economy is defined by Isaak as 'the study of the patterns of positioning and collective learning between nations and peoples that either preserve or change the inequality between them' (1995:298).

8   I take issue with Murphy's general thesis of community-based approach to tourism planning and development on the grounds that it is a North American model which assumes levels of democracy, education, knowledge and power sharing that simply doesn't exist in many countries.

9   The basic needs philosophy dominated the International Labour Organisation, United Nations and World Bank during the decade of the 1970s and beyond. It was a backlash against the purely economic measurement of growth as an alternative to GNP or GDP per capita

measure. It focused on the provision of health services, education, housing, sanitation, water and adequate nutrition so as to improve the quality of life for the poorest people.

10 This rivalry places many Third World countries at a particular disadvantage, for example 'unless nations possess large quantities of resources, like raw materials, capital, technology and the like, the tactic of closing off your territory to [multinational corporations and foreign direct investment] in today's world is a little like shooting yourself in the foot' (Balaam and Veseth, 1996:352).

11 However, Europe had critical advantages over colonial countries or former colonies. For example, even though the European countries were exhausted and bankrupt, they possessed the necessary structural, institutional and cultural conditions to make aid work: well-integrated commodity and money markets; highly developed transport technologies; well trained, educated and experienced labour; the motivation to succeed; very experienced governments and a history of having experienced complex capitalist development within living memory.

12 Todaro describes these as 'a newly established industry, usually set up behind the protection of a tariff barrier as part of a policy of import substitution. Once the industry is no longer an infant, the protective barriers are supposed to disappear, but often they do not' (1997:699).

Bibliography

Abraham, J. (1973) *The Origins and Growth of Sociology*, Harmondsworth: Penguin.

Adams, W. M. (1990) *Green Development: Environment and Sustainability in the Third World*, London: Routledge.

Aziz, H. (1995) 'Understanding Terrorist Attacks on Tourists in Egypt', *Tourism Management*, 16 (2): 91–7.

Balaam, D. and Veseth, M. (1996) *Introduction to the International Political Economy*, New Jersey: Prentice-Hall.

Baran, P. (1967) *The Political Economy of Growth*, New York: Monthly Review Press (orginally published in Spanish in 1957).

Barley, N. (1983) *The Innocent Anthropologist: Notes from a Mud Hut*, Harmondsworth: Penguin.

Barthes, R. (1984) *Mythologies*, London: Paladin.

Bauer, P. (1981) *Equality, the Third World and Economic Delusion*, London: Weidenfeld and Nicolson.

Bayley, S. (1991) *Taste: the Secret Meaning of Things*, London: Faber and Faber.

Boissevain, J. (ed.) (1996) *Coping With Tourists: European Reactions to Mass Tourism*, Providence: Berghahn Books.

Boniface, P. and Fowler, P. (1993) *Heritage and Tourism in 'the Global Village'*, London: Routledge.

Boorstin, D. (1964) *The Image: a Guide to Pseudo-Events in America*, New York: Harper and Row.

Bowen, J. and Basch, J. (1992) 'Strategies for Creating Customer-Oriented Organizations', (in) Teare, R. and Olsen, M. (1992).

Britton, S. (1982) 'The Political Economy of Tourism in the Third World', *Annals of Tourism Research*, 9: 331–58.

Brown, D. (1996) 'Genuine Fakes', (in) Selwyn, T. (1996).

Bryden, J. (1973) *Tourism and Development: a Case Study of the Commonwealth Caribbean*, Cambridge: Cambridge University Press.

Buck, R. (1978) 'Towards a Synthesis in Tourism Theory', *Annals of Tourism Research*, 1: 110–111.

Burns, P. and Cleverdon, R. (1995) 'Destination on the Edge? The case of the Cook Islands' (in) Conlin, M. and Baum, T. (eds).

Burns, P. and Cleverdon, R. (1999) 'Planning Tourism in a Restructuring Economy: the Case of Eritrea', (in) Dieke, P. (ed.) *The Political Economy of Tourism in Africa*, New York: Cognizant.

Burns, P. and Holden, A. (1995) *Tourism: a New Perspective*, Englewood Cliffs: Prentice-Hall.

Cheater, A. (1989) *Social Anthropology: an Alternative Introduction*, London: Routledge.

Cohen. E. (1971) 'Arab Boys and Tourist Girls in a Mixed Arab / Jewish Community', *International Journal of Comparative Sociology*, XII (4): 217–33.

—— (1972) 'Towards a Sociology of International Tourism', *Social Research*, 39 (1).

—— (1974) 'Who is a Tourist?: A Conceptual Classification', *Sociological Review*, 22: 527–55.

—— (1988) 'Authenticity and Commoditization in Tourism', *Annals of Tourism Research*, 15: 371–86.

Conlin, M. and Baum, T. (eds) (1995) *Island Tourism: Management Principles and Practice*, Chichester: Wiley.

Cooper, C., Fletcher, J., Gilbert, D. and Wanhill, S. (1993) *Tourism Principles and Practice*, London: Longman.

Corner, J. and Harvey, S. (1991) *Enterprise and Heritage: Crosscurrents of National Culture*, London: Routledge.

Court, W. H. B. (1967) *A Concise Economic History of Britain: from 1750 to Recent Times*, London: CUP.

Crick, M. (1989) 'Sun, Sex, Sights, Savings and Servility: Representations of Tourism in the Social Sciences', *Criticism, Heresy and Interpretation*, I (1), 37–76

Cypher, J. and Dietz, J. (1997) *The Process of Economic Development*, London: Routledge.

Dann, G. (1996) 'The People of Tourist Brochures', (in) Selwyn, T. (1996).

—— (1997) *The Language of Tourism*, Oxford: Pergamon.

Deitch, L. (1977) 'The Impact of Tourism on the Arts and Crafts of the Indians of the Southwestern United States' (in) Smith, V. (1977).

de Kadt, E. (1992) 'Making the Alternative Sustainable: Lessons from Development for Tourism', (in) Smith, V. and Eadington, W. (1992).

Doray, B. (1988) *From Taylorism to Fordism: a Rational Madness*, London: Free Association Books.

Doxey, G. (1975) 'A Causation Theory of Visitor–Resident Irritants: Methodology and Research Inferences' (in) *Proceedings of the Travel Research Association* 6th Annual Conference, San Diego, California.

—— (1976) 'When Enough's Enough: the Natives are Restless in Old Niagara', (cited in) Mathieson, A. and Wall, G. (1982).

Dunning, J. (1993) *The Globalization of Business: the Challenge of the 90s*, London: Routledge.

During, S. (ed.) (1993) *The Cultural Studies Reader*, London: Routledge.

Durkheim, E. (1915 [1965]) *Elementary Forms of Religious Life*, London: Free Press.

Edgell, David L. (1990) *International Tourism Policy*, New York: Van Nostrand Reinhold.

Enloe, C. (1989) *Bananas, Beaches and Bases: Making Feminist Sense of International Politics*, London: Pandora.

Farrell, B. (1982) *Hawaii, the Legend that Sells*, Honolulu: University of Hawaii Press.

Frank, A.G. (1966) 'The Development of Underdevelopment', *Monthly Review*, xviii, 14.

Friberg, M. and Hettne, B. (1985) 'The Greening of the World: Towards a Non-Deterministic Model of Global Processes' (cited by) Adams, W. M. (1990).

Gee, C., Makens, J. and Choy, D. (1989) *The Travel Industry* (2nd edition), New York: Van Nostrand Reinhold.

George, S. (1989) *A Fate Worse than Debt*, London: Penguin.

Ghatak, S. (1995) *Introduction to Development Economics* (3rd edition), London: Routledge.

Giddens, A. (1990) *The Consequences of Modernity*, Cambridge: Polity Press.

Gluckman, M. (1955) *Customs and Conflicts in Africa*, Oxford, Blackwells.

Go, F. and Pine, R. (1995) *Globalization Strategy in the Hotel Industry*, London: Routledge.

Goulet, D. (1971) *The Cruel Choice: a New Concept on the Theory of Development*, New York: Atheneum.

Graburn, N. (1977) 'Tourism: The Sacred Journey' (in) Smith, V. (ed.) (1977).

—— (1983) 'To Pray, Pay and Play: the Cultural Structure of Japanese Domestic Tourism', *Les Cahiers du Tourisme*, serie no. 26, Centre des Haute Etudes Touristiques, Université de Droit, E'Economie et des Sciences, Aix-en-Provence: Centre des Hautes Etudes Touristiques.

—— (1984) 'The Evolution of Tourist Arts', *Annals of Tourism Research*, 11: 393–419.

—— (1989) 'Tourism: the Sacred Journey', (in) Smith, V. (1989).

Greenwood, D. (1989) 'Culture by the Pound: an Anthropological Perspective on Tourism as Cultural Commoditization', (in) Smith, V. (1989).

Hall, C. M. (1994) *Tourism and Politics*, Chichester: Wiley.

Harrison, Paul (1993) *Inside the Third World* (3rd edition), London: Penguin.

Haviland, W. (1990) *Cultural Anthropology* (6th edition), Fort Worth: Holt, Rinehart and Winston Inc.

Hiller, S. (1991) *The Myth of Primitivism*, London: Routledge.

Hitchcock, M., King, V. and Parnwell, M. (eds) (1993) *Tourism in South East Asia*, London: Routledge.

Hobart, M. (ed.) (1993) *An Anthropological Critique of Development: the Ignorance of Growth*, London: Routledge.

Hofstede, G. (1991) *Cultures and Organizations: Softwares of the Mind*, Maidenhead: McGraw-Hill.

Hoggart, R. (ed.) (1992) *The Oxford Illustrated Encyclopedia of Peoples and Cultures*, Oxford: Oxford University Press.

Holloway, C. (1994) *The Business of Tourism* (4th edition), London: Longman.

Hoogvelt, A. (1982) *The Third World in Global Development*, Basingstoke: Macmillan.

—— (1997) *Globalisation and the Postcolonial World: the New Political Economy of Development*, Basingstoke: Macmillan.

Howard, M. (1996) *Contemporary Cultural Anthropology* (5th edition), New York: HarperCollins.

Isaak, R. (1995) *Managing World Economic Change: International Political Economy*, Englewood Cliffs: Prentice-Hall.

Jafari, J. (1977) Editor's page, *Annals of Tourism Research*, V (i).

Jones, P. (1993) 'Operations Management Issues', (in) Jones, P. and Pizzam, A. (eds).

Jones, P. and Pizzam, A. (eds) (1993) *The International Hospitality Industry: Organizational and Operational Issues*, London: Pitman.

Kay, C. (1989) *Latin American Theories of Development and Underdevelopment*, London: Routledge.

Keesing, R. and Keesing, F. (1971) *New Perspectives in Cultural Anthropology*, New York: Holt, Rinehart and Winston Inc.

Kuper, A. (1988) *The Invention of Primitive Society: Transformations of an Illusion*, London: Routledge.

Laws, E. (1991) *Tourism Marketing: Service and Quality Management Perspectives*, Cheltenham: Stanley Thornes.

Lea, J. (1988) *Tourism and Development in the Third World*, London: Routledge.

Leiper, N. (1995) *Tourism Management*, Collingwood: TAFE Publications.

Lévi-Strauss, C. (1963) *Structural Anthropology*, New York: Basic Books.

Lodge, D. (1991) *Paradise News*, Harmondsworth: Penguin.

MacCannell, D. (1976) *The Tourist*, New York: Schocken.

—— (1992) *Empty Meeting Ground: the Tourist Papers*, London: Routledge.

McKean, P. (1977) 'Towards a Theoretical Analysis of Tourism: Economic Dualism and Cultural Introversion in Bali' (in) Smith, V. (ed.) (1977).

McLuhan, M. (1960) *Explorations in Communication*, Boston: Beacon Press.

Malinowski, B. (1922) *Argonauts of the Western Pacific*, London: Routledge and Kegan Paul.

Mathieson, A. and Wall, G. (1982) *Tourism: Economic, Physical and Social Impacts*, Harlow: Longman.

Mayle, P. (1989) *A Year in Provence*, London: Hamilton.

Meek, R. (1976) *Social Science and the Ignoble Savage*, Cambridge: CUP.

Mehmet, O. (1995) *Westernizing the Third World: the Eurocentricity of Economic Development Theories*, London: Routledge.

Middleton, V. (1998) *Marketing in Travel and Tourism*, Oxford: Heinemann.

Mill, R. and Morrison, A. (1985) *The Tourism System*, New Jersey: Prentice-Hall.

Murphy, P. (1985) *Tourism: a Community Approach*, London: Routledge.

Nash, D. (1977) 'Tourism as a Form of Imperialism', in Smith, V. (ed.) (1989[1977]).

—— (1981) 'Tourism as an Anthropological Subject', *Current Anthropology*, 22 (5), 461–81.

—— (1989) 'Tourism as a Form of Imperialism', (in) Smith, V. (ed.) (1989[1977]).

—— (1996) *The Anthropology of Tourism*, Oxford: Elsevier.

Nicholson-Lord, D. (1990) 'Death by Tourism', *Independent on Sunday*, 5 August.

Nuñez, T. (1989) 'Touristic Studies in Anthropological Perspective', (in) Smith, V. (ed.) (1989).

Olsen, M. and Merna, K. (1993) 'The Changing Nature of the Multinational Hospitality Firm', (in) Jones, P. and Pizzam, A. (1993).

Passariello, P. (1983) 'Never on a Sunday? Mexican Tourists at the Beach', *Annals of Tourism Research*, 10: 109–22.

Pattullo, P. (1996) *Last Resorts: the Cost of Tourism in the Caribbean*, London: Cassell.

Pearce, P. (1982) *The Social Psychology of Tourist Behaviour*, Oxford: Pergamon.

Peet, R. (1991) *Global Capitalism: Theories of Societal Development*, London: Routledge.

Pepper, D. (1993) *Eco-Socialism: from Deep Ecology to Social Justice*, London: Routledge.

Pfaffenberger, B. (1983) 'Serious Pilgrims and Frivolous Tourists', *Annals of Tourism Research*, 10: 57–74.

PKF (1993) *Trends in the Hotel Industry*, London: Pannell Kerr Foster.

Plog, S. (1977) 'Why destinations rise and fall in popularity' in ICTA *Domestic and International Travel*, Massachusetts: Wellesley.

Plog, S. (1994) 'Leisure Travel: an Extraordinary Industry Faces Superordinary Problems', (in) Theobold, W. (ed.) (1994).

Poon, A. (1993) *Tourism, Technology and Competitive Strategies*, Wallingford: CAB.

Pottier, J. (ed.) (1993) *Practising Development: Social Science Perspectives*, London: Routledge.

Radcliffe-Brown, A. (1922) *The Andaman Islanders*, Cambridge: Cambridge University Press.

Rajotte, F. (1980) 'Introduction', (in) Rajotte, F. and Crocombe, R. (eds) *Pacific Tourism, as Islanders See it*, Suva: USP.

Richter, L. (1989) *The Politics of Tourism in Asia*, Honolulu: University of Hawaii Press.

Ritzer, G. (1993) The *McDonaldization of Society*, Newbury Park: Pine Forge Press.

—— (1998) *The McDonaldization Thesis: Explorations and Extensions*, London: Sage.

Robertson, R. (1992) *Globalization: Social Theory and Global Culture*, London: Sage.

Rostow, W. W. (1960) *The Stages of Economic Growth, a Non-Communist Manifesto*, Cambridge: CUP.

—— (1978) *Getting from Here to There*, London: Macmillan.

Ryan, C. (1991) *Recreational Tourism: a Social Science Perspective*, London: Routledge.

Saakana, A. (1988) 'Mythology and History: An Afrocentric Perspective of the World' *Third Text*, 3/4: 143–50.

Said, E. (1978) *Orientalism: Western Concepts of the Orient*, London: Routledge and Kegan Paul.

Seaton, A. and Bennett, M. (1996) *The Marketing of Tourism Products: Concepts, Issues and Cases*, London: International Thomson Business Press.

Selwyn, T. (1993) 'Peter Pan in South East Asia', (in) Hitchcock, M. *et al.* (eds) (1993).

Selwyn, T. (1994) 'The Anthropology of Tourism: Reflections on the State of the Art' (in) Seaton, A. *et al.* (eds) (1994) *Tourism: the State of the Art*, London: Wiley.

—— (1996) (ed.) *The Tourist Image: Myths and Myth Making in Tourism*, London: Wiley.

Sewell, B. (1996) Monumental Destruction *Arts Review*, September (cited in) *Tourism in Focus*, 22 Winter 1996/7.

Shaw, G. and Williams, A. (eds) (1997) *The Rise and Fall of British Coastal Resorts: Cultural and Economic Perspectives*, London: Pinter.

Shields, R. (1992) *Lifestyle Shopping: the Subject of Consumption*, London: Routledge.

Smith, V. (1981) response to Nash, D. 'Tourism as an Anthropological Subject', *Current Anthropology*, 22 (5): 461–81.

—— (ed.) (1977) *Hosts and Guests: the Anthropology of Tourism* (1st edition), Philadelphia: University of Pennsylvania Press.

—— (ed.) (1989) *Hosts and Guests: the Anthropology of Tourism* (2nd edition), Philadelphia: University of Pennsylvania Press.

Smith, V. and Eadington, W. (1992) *Tourism Alternatives: Potentials and Problems in the Development of Tourism*, Philadelphia: University of Pennsylvania Press.

Spicer, E. (1968) 'Acculturation', (in) Sill, D. (ed.) *International Encyclopaedia of the Social Sciences*, (vol. 3, pp. 21–5).

Teare, R. and Olsen, M., (eds) (1992) *International Hospitality Management*, Chichester: Pitman.

Theroux, P. (1992) *The Happy Isles of Oceania: Paddling the Pacific*, London: Hamish Hamilton.

Thirlwall, A. (1989) *Growth and Development with Special Reference to Developing Economies*, Basingstoke: Macmillan.

Todaro, M. (1982) *Economics for a Developing World* (2nd edition), Harlow: Longman.

—— (1997) *Economic Development* (6th edition), Harlow: Longman.

Toye, J. (1993) *Dilemmas of Development*, Oxford: Blackwell.

Turner, B. (1994) *Orientalism, Post-Modernism and Globalism*, London: Routledge.

Turner, L. and Ash, J. (1975) *The Golden Hordes: International Tourism and the Pleasure Periphery*, London: Constable.

Turner, V. (1967) *The Forest of Symbols*, Ithaca: Cornell University Press.

Turner, V. and Turner, E. (1978) *Image and Pilgrimage in Christian Culture: Anthropological Perspectives*, New York: Columbia University Press.

Tylor, E. Burnett (1871 [1924]) *Primitive Culture* (7th edition), New York: Brentano's.

UNDP (1990) *United Nations Human Development Index*, New York: UNDP.

—— (1993) *Guidelines for Project Formulation and the Project Document Format*, New York: UNDP.

—— (1995) *Human Development Report*, New York: UNDP.

Urry, J. (1990) *The Tourist Gaze*, London: Sage Publications.

Uzzell, D. (1984) 'An Alternative Structuralist Approach to the Psychology of Tourism Marketing', *Annals of Tourism Research*, 11: 79–99.

van Gennep, A. (1960 [1908]) *The Rites of Passage*, Chicago: University of Chicago Press.

van Harssel, J. (1994) *Tourism: an Exploration* (3rd edition), Englewood Cliffs: Prentice-Hall.

Wahab, S. (1975) *Tourism Management*, London: Tourism International Press.

Waldren, J. (1996) *Insiders and Outsiders: Paradise and Reality in Mallorca*, Oxford: Berghahn Books.

Waters, M. (1995) *Globalization*, London: Routledge.

Witt, S. and Moutinho, L. (eds) (1995) *Tourism Marketing and Management Handbook*, Hemel Hemstead: Prentice-Hall.

Wood, R. (1993) 'Tourism, Culture and the Sociology of Development', (in) Hitchcock, M. *et al.* (eds) (1993).

Wolff, K. (1950) (Editor and translator) *The Sociology of Georg Simmel*, London: The Free Press.

WTO (1998) *Tourism Highlights, 1997*, Madrid: World Tourism Organization.

Young, G. (1973) *Tourism: Blessing or Blight?*, Harmondsworth: Penguin.

Index

Page numbers in *italics* refer to figures and tables